BASIC STUDY OF ORGANIC CHEMISTRY

BASIC STUDY
OF
ORGANIC CHEMISTRY

Amar Tyagi

ANMOL PUBLICATIONS PVT. LTD.
NEW DELHI - 110 002 (INDIA)

ANMOL PUBLICATIONS PVT. LTD.
H.O.: 4374/4B, Ansari Road, Darya Ganj,
New Delhi-110 002 (India)
Ph.: 23261597, 23278000
B.O.: No. 1015, Ist Main Road, BSK IIIrd Stage
IIIrd Phase, IIIrd Block,
Bangalore - 560 085 (India)
Visit us at: www.anmolpublications.com

Basic Study of Organic Chemistry

First Published, 2006

ISBN 81-261-2670-1

PRINTED IN INDIA

Published by J.L. Kumar for Anmol Publications Pvt. Ltd., New Delhi - 110 002 and Printed at Mehra Offset Press, Delhi.

CONTENTS

PREFACE

This book deals with the significance of atoms in the organic chemistry. All possible efforts have been made to enhance further the usefulness of the book. This book covers all the topics of organic chemistry.

This book contains topics on electrons, configuration of atoms, molecular orbital, resonance etc. and other topic can be seen from the contents. The other highlights of the book are: (i) the subject contents have been developed, refined and reconstructed at several points, (ii) several new exercises have also been added at the end of various chapters for the benefit of students. (iii) every page of the book has been read very carefully so as to improve its quality.

I am grateful to all those who helped me directly or indirectly in preparing this edition. I firmly believe that there is always scope for improvement and accordingly I shall look forward to receive suggestions, (which shall be thankfully acknowledged) to further enrich the quality of the text.

Author

Chapter 1
ELECTRONIC CONFIGURATION OF ATOMS

THE ATOM

Atom may be defined as the smallest particle of an element that participates in a chemical reaction. The center of the atom consists of nucleus which is responsible for almost, the entire mass of the atom. The nucleus is surrounded by electrons which is equal to the number of protons in the nucleus. Nucleus also contains one more particle named neutrons.

The mass of a proton has been found to be $1.6726485 \times 10^{-24}$g or 1.0072764 *amu* and carries a charge of $+ 1.60 \times 10^{-19}$ couloumb, called a unit positive charge (+1). A neutron has a mass of $1.6749543 \times 10^{-24}$ g or 1.0086650 *amu*, which is almost the same as the mass of the proton and has no charge. Nucleus is thus positively charged and its net positive charge is equal to the number of protons in it. This number is characteristic of cach element and is known as the atomic number of the element.

Electron is a negatively charged particle having a charge of -1.60×10^{-19} coulomb, called a unit negative charge (-1). The mass of an electron is 9.109534×10^{-28}g or 5.485802×10^{-4} *amu* which is 1/1837of the mass of a hydrogen atom. Since the number of electrons surrounding the nucleus of an atom is equal to the number of protons present in the nucleus, so the atom becomes electrically neutral. The electron spins about its axis and revolves around the nucleus. The physical and chemicel properties of on atom and that of an element depend upon the number, the arrangement and the behaviour of the extra- nuclear electrons, under different conditions.

THE QUANTUM NUMBERS

(1) **Particle nature of waves:-** In 1900 Max Planck introduced the concept that energy is quantized i.e., quantum theory of radiation. Planck's equation is E=*hv*, where E is energy, 'h' is Planck's constant and *v*, the frequency. The value of *hv* gives the size of a quantum of radiation. Planck stated a hot body emits radiant energy discontinuously in the form of small packets of energy called Quantum (plural is Quanta). The energy emitted or absorbed can be either one quantum or any whole multiple of *hv* i.e. *2hv, 3hv, 4v.....nhv* quanta of energy.

Wave nature of Particle. L. de Broglie in 1923 proposed that if radiation has particle-like properties, then particles in motion should have wave-like properties. He predicted that the wave-lengh λ of a moving particle can be calculated from the equation λ = h/mv, where h is Planck's constant, m, the mass of the particle and v, the velocity of the particle. In 1927, C.J. Davisson and L.H. Germer observed the *diffraction* of an electronbeam. The wavelength of the electron-beam calculated from the diffraction pattern is in equivalence with the wavelength calculated from Broglie's equation. Thus it is seen that the electrons too behave like waves. It should be borne in mind that the wave properties of matter are significant only for particles of very small mass. An electron, thus, behaves both as a particle and as a wave depending upon the conditions of the experiment, *but can not behave simultaneously as both.* In 1926, two German physicists, W. Heisenberg and E. Schrodinger studied the mathematics of the wave properties of the electrons in atoms and gave the theory of wave-mechanics, description of which is beyond the scope of our discussion. Using Schrodinger's equation we can compute the energy level of an electron and probability of finding it in a particular region surrounding the nucleus. The space in which an electron with a specific energy is most likely to be found the *energy level* for a particular electron-is called an **atomic orbital**. *Heisenberg's uncertainty principle* states that it is impossible to know simultaneously both the exact momentum and exact location of an electron. Hence, it is understood from the principle that the electron is somewhere within the orbital but we can't know its exact location.

To search how electrons are distributed in atoms of the different elements identification numbers are required to specify a particular electron in an atom. These identification numbers are called *Quantum numbers*. These are

(A) Principal Quantum Numbers

(B) Azimuthal Quantum Number or Angular Momentum Q.N

(C) Magnetic Quantum Number

(D) Spin Quantum Number

Principal Quantum number. Different energy levels are expressed in terms of the numbers (1,2 3, 4,....) counted from the nucleus. These numbers are known as principal quantum numbers represented by n. Electrons having having greater values of 'n' have greatest energy and are far away from the nucleus than those with lower number. Values of *n* from 1 to 7 are known. These energy levels or shells are also denoted by the letters K,L,M N, etc. respectively in place of the numbers. These letters are derived from the names of the groups of lines present in the X-ray spectra characteristic of the positions of the electrons in these energy levels.

$2n^2$ is the maximum number of electron which can be contained by each shell, where n is the principal quantum numer of the shell. For example—first energy level, or K shell, where n=1, has the maximum capacity of 2×1^2 i.e., 2 electrons; second energy level, or L shell, where n=2 has the maximum capacity of 2×2^2 *i.e.* 8 electrons; third energy level; or M shell, where n=3, has the maximm capacity of 2×3^2 *i.e.* 18 electrons, and N shell or 4th energy level whose n=4 can contain maximum of $2 \times 4^2 = 32$ electrons.

Angular momentum quantum nmber. Also known as Azimuthal Quantum Number. Electrons within each principal energy level occupy different energy levels called subshells, or sublevels. Angular momentum Quantum number describes spatial distribution of electron cloud and angular momentum of the electrom It tells. the shape of the orbital occupied by the electron Each value corresponds to on energy subshell or subsell. The number of possible subshells increases with n. The possible values of l are 0 to n-1. Thus, the K shell with n=1 has one l, level =0, L shell with n=2 has l=0 and 1, M shell with n=3 has l=0,1 and 2, and N shell with n=4 has I=0,1,2 and 3. The first four subshells with angular momentum quantum numbers 0,1,2 and 3 are designated by the letters s, p, d and f respectively. The letters s, p, d, e, f, are taken from the words-sharp, principal, diffuse and fundamental which are used to describe the spectral series connected with the corresponding l values. There may be at the most electrons in s,6 in p, 10 in d and 14 in f sub-shell.

Thus—

K shell, where n=1and l=0, has only one subshell, written as 1s. The two electrons of K shell occupy 1s subshell.

L shell, where n= 2 and l=0 and 1, has two subshells written as 2s and 2p. Two out of eight electrons of this shell are in 2s and the remaining six in 2p.

M shell, where n=3 and l=0, 1 and 2, has *three subshells,* written as 3*s*, 3*p* and 3*d*. Out of the 18 electrons of this shell, two are in 3*s*, six in 3*p* and ten in 3*d*.

N-shell n=4 and n=4 and l=0,1,2 and 3 has four subshells which are 4s, 4p, 4d and 4f. Out of 32-e in N-shell 2, 6, 10, 14, recide in 4s, 4p, 4d and 4f respectively.

Magnetic Quantum number. It has been found that *s* subshell is spherically symmetrical whereas, *d* and *f* subshells exist in more than one direction, viz., *p* exists in three, *d* in five and *f* in seven different directions. The orbitrary regions in the space where there is probability of finding electrons of these subshells are termed as atomic orbitals. These directions create different magnetic effects which are distinguished by third quantum number, called magnetic quantum number denoted by m. The values of m range from $-l$ to$+l$ (increasing by unity). The energy of atomic orbitals in a particular subshell is the same. Thus—

S-subshell, l=0, since l=0, so the magnetic quantum number does not exist and has a single spherically symmetrical orbital.

psubshell, whose l=1, has m=−1, 0,+1 *i.e. three p atomic orbitals* written as p_x, p_y and p_z lying along three normal axes x, y and z.

d subshell, whose l=2, has m−2−1, 0,+1, +2 *i.e. five d atomic orbitals* written as d_1 ,d_2, d_3, d_4 and d_5 lying in five different directions.

f subshell, whose l=3 has m= -3,-2,-1,0,+1,+2,+3 *i.e. seven f atomic orbitals* written as f_1,f_2, f_3,f_4, f_5,f_6, f_7 in seven different directions.

In any energy level, n^2 is the total number of atomic orbitals present, n is principal quantum number. Hence,

In K shell, n=1,number of orbitals is l *i.e.* 1s.

In L shell, n=2, number of orbitals is 4 *i.e.* 2s, $2p_x$, $2p_y$,$2p_z$

In M shell, n=3, number of orbitals is 9 *i.e.* 3s, $3p_x$, $3p_y$, $3p_z$, $3d_1$, $3d_2$, $3d_3$, $3d_4$, $3d_5$

In N shell, n=4 number of orbitals is 16 *i.e.* 4s, $4p_x$, $4p_y$, $4p_z$, $4d_1$, $4d_2$, $4d_3$, $4d_4$, $4f_1$, $4f_2$,$4f_3$, $4f_4$, $4f_5$, $4f_6$, $4f_7$.

Any orbital can contain maximum of 2 electrons.

Spin quantum number. The two electrons of the same orbital differ in the direction of their spin-clockwise and anticlockwise. In this state of antiparallel spin, the two electrons of the same orbital are said to be paired and graphically represented as ↑↓ (arrows stand for electrons and their heads for the directions of their spins) but spinning in opposite direction *i.e.* one clockwise and other anticlockwise are recognised by the fourth quantum number called spin quantum number whose values are +1/2 and −1/2.

Pauli exclusion principle. All the above mentioned facts are summarised in fundamental *Pauli exclusion principle,* which states that no two electrons in an atom can have the same set of four quantum numbers.

ELECTRON-DISTRIBUTION IN ATOMIC ORBITALS

Keep the following points in mind while distributing electrons in different orbitals of an atom of a given element:

1. *Electrons tend to occupy the lowest energy orbitals.* The order of the increasing energy content in different orbitals is as follows:

1s < 2s < 2p < 3s < 3p < 4s <3d < 4p < 5s < 4d < 5p < 6s and so on (Fig. 1)

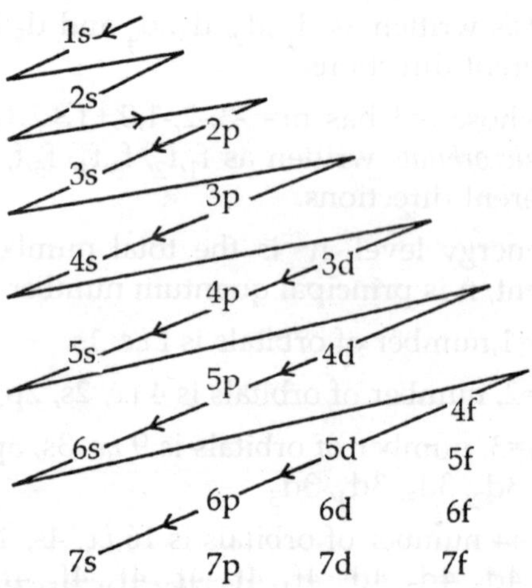

Fig. 1 – Diagram showing the order of increasing energy content of the orbitals.

2. **Hund's Rule:**

(a) Electron pairing in p, d, and f orbitals cannot occur until each orbital of a given subshell contains one electron each or is singly occupied

(b) Electrons present singly in different orbitals of a given subshell tend to have parallel spins.

The distribution of electrons in orbitals and the electronic configuration of the first eighteen elements is given in the charts on pages 7and 8 respectively.

CONDENSED FORM OF THE MODE OF ELECTRON-DISTRIBUTION IN ORBITALS

Main energy shell with principal quantum number	K n = 1	L n = 2		M n = 3			N n = 4			
Maximum number of electrons in the energy shell	2	8		18			32			
Subshells	s	s	p	s	p	d	s	p	d	f
Maximum number of electrons in the subshells	2	2	6	2	6	10	2	6	10	14
Number of orbitals in the corresponding subshells	1	1	3	1	3	5	1	3	5	7
Names of the orbitals (2 electrons in each orbital with antiparallel spins represented by ↑ ↓)	s ↑ ↓	s ↑ ↓	p_x ↑ ↓ p_y ↑ ↓ p_z ↑ ↓	s ↑ ↓	p_x ↑ ↓ p_y ↑ ↓ p_z ↑ ↓	d_1 ↑ ↓ d_2 ↑ ↓ d_3 ↑ ↓ d_4 ↑ ↓ d_5 ↑ ↓	s ↑ ↓	p_x ↑ ↓ p_y ↑ ↓ p_z ↑ ↓	d_1 ↑ ↓ d_2 ↑ ↓ d_3 ↑ ↓ d_4 ↑ ↓ d_5 ↑ ↓	f_1 ↑ ↓ f_2 ↑ ↓ f_3 ↑ ↓ f_4 ↑ ↓ f_5 ↑ ↓ f_6 ↑ ↓ f_7 ↑ ↓
Total number of orbitals in main shell $=n^2$	1	4		9			16			

Note - To write the position of electrons in atomic orbitals, following method is adopted:

Number of main energy level | Name of subshell | Number of electrons in that orbital. | Direction or number of the orbital.

e.g. the two electrons of p_y orbital of L shell will be written as $2p^2_y$

CHART SHOWS THE ELECTRONIC CONFIGURATION OF EIGHTEEN ELEMENTS

Atomic No.	Element	Electronic configuration	K shell n = 1	L shell (n = 2)				M shell (n = 3)			
					2p (l = 1)				3p (l = 1)		
			1s l = 0 m = 0	2s l = 0 m = 0	2px m = -1	2py m = 0	2pz m = +1	3s l = 0 m = 0	3px m = -1	3py m = 0	3pz m = +1
1	H	$1s^1$	↑								
2	He	$1s^2$	↑↓								
3	Li	$1s^2 2s^1$	↑↓	↑							
4	Be	$1s^2 2s^2$	↑↓	↑↓							
5	B	$1s^2 2s^2 2p_x^1$	↑↓	↑↓	↑						
6	C	$1s^2 2s^2 2p_x^1 2p_y^1$	↑↓	↑↓	↑	↑					
7	N	$1s^2 2s^2 2p_x^1 2p_y^1 2p_z^1$	↑↓	↑↓	↑	↑	↑				
8	O	$1s^2 2s^2 2p_x^2 2p_y^1 2p_z^1$	↑↓	↑↓	↑↓	↑	↑				
9	F	$1s^2 2s^2 2p_x^2 2p_y^2 2p_z^1$	↑↓	↑↓	↑↓	↑↓	↑				
10	Ne	$1s^2 2s^2 2p_x^2 2p_y^2 2p_z^2$	↑↓	↑↓	↑↓	↑↓	↑↓				
11	Na	$1s^2 2s^2 2p_x^2 2p_y^2 2p_z^2 3s^1$	↑↓	↑↓	↑↓	↑↓	↑↓	↑			
12	Mg	$1s^2 2s^2 2p_x^2 2p_y^2 2p_z^2 3s^2$	↑↓	↑↓	↑↓	↑↓	↑↓	↑↓			
13	Al	$1s^2 2s^2 2p_x^2 2p_y^2 2p_z^2 3s^2 3p_x^1$	↑↓	↑↓	↑↓	↑↓	↑↓	↑↓	↑		
14	Si	$1s^2 2s^2 2p_x^2 2p_y^2 2p_z^2 3s^2 3p_x^1 3p_y^1$	↑↓	↑↓	↑↓	↑↓	↑↓	↑↓	↑	↑	
15	P	$1s^2 2s^2 2p_x^2 2p_y^2 2p_z^2 3s^2 3p_x^1 3p_y^1 3p_z^1$	↑↓	↑↓	↑↓	↑↓	↑↓	↑↓	↑	↑	↑
16	S	$1s^2 2s^2 2p_x^2 2p_y^2 2p_z^2 3s^2 3p_x^2 3p_y^1 3p_z^1$	↑↓	↑↓	↑↓	↑↓	↑↓	↑↓	↑↓	↑	↑
17	Cl	$1s^2 2s^2 2p_x^2 2p_y^2 2p_z^2 3s^2 3p_x^2 3p_y^2 3p_z^1$	↑↓	↑↓	↑↓	↑↓	↑↓	↑↓	↑↓	↑↓	↑
18	Ar	$1s^2 2s^2 2p_x^2 2p_y^2 2p_z^2 3s^2 3p_x^2 3p_y^2 3p_z^2$	↑↓	↑↓	↑↓	↑↓	↑↓	↑↓	↑↓	↑↓	↑↓

S AND P ORBITALS-SHAPES

An electron moving very fast can be found all over the space it covers, with the charge stretched throughout. The electron lossesits particle nature and there is only electron charge cloud existing in the space travelled by the electron around the nucleus. Mathematically, electrons can be assigned a wave-function ψ and the probability of finding an electron in the element of volume dv can be known by ψ^2 dv. Hence we can find the shapes of these orbitals of different sub-energy levels.

The s atomic orbital of any energy level is spherical with nucleus at the centre (Fig. 2A). On account of this spherical shape,s orbitals can make a bond equally well in all directions, The probability of finding p-electrons is maximum in two lobes on the opposite sides of the nucleus.This gives rise to a dumbbell shape for the p- orbital. However. note that the probability to find particular p- electronis equal in both the lobes.The two lobes are seperated by a nodal- plane where probability of bonding electron is zero.The three dumb-bells representing the three p orbitals are oriented along the axis x, y and z perpendicular to each other i.e. each p orbital has an axis at right angles to the other two (Fig. 2B, 2C and 2D).

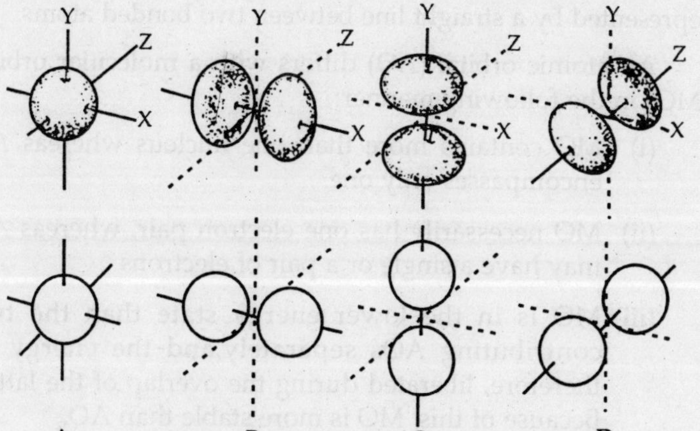

Fig.2—Shapes of the atomic orbitals,
A—s; B—px; C—py; D—yz

Chapter 2

MOLECULAR ORBITALS

When two atoms come so close that their atomic orbitals overlep to form a new single orbital including the nuclei of the two atoms is termed as Bonding and the new orbital formed is known as molecules orbital. The electrons present singly in the atomic orbitals now get paired (with opposite spins) to be known as **bonding electron pair**, which resides in the newly formed molecular orbital. The electron-pair makes the covalent bond between the two atoms and this bond is represented by a straight line between two bonded atoms.

An atomic orbital (AO) differs with a molecular orbital (MO) in the following manner:

(i) MO contains more than one nucleus whereas AO encompasses only one.

(ii) MO necessarily has one electron pair, whereas AO may have a single or a pair of electrons.

(iii) MO is in the lower energy state than the two contributing AOs separately,and the energy is, therefore, liberated during the overlap of the latter. Because of this, MO is more stable than AO.

(iv) MO are represented by σ, σ*, π, π*,etc. They may be bonding or ontibunding AO are represented by s,p,d,f, etc.

The greater stability of MOs is due to the greater force of attraction on their electrons by the combined effect of two nuclei, as compared to the force of attraction exercised by the single nucleus on the electrons of AOs.

Sigma (σ) molecular orbital

Coaxial (along same axes) overlap of atomic orbitals results molecular orbital termed as Sigma σ moleculer orbitals. The electrons resinding in this MO are called σ electrons, and the bond so formed is called a σ bond. For example

Overlap: of s-s orbitals: Here the bonding molecular orbital has been designated as σ (is) for 15 overlep with is and σ (25) for 25 overlapping 25. When two hydrogen atoms come very close to each other, their 1s orbitals overlap to form a new σ molecular orbital (Fig.3.)

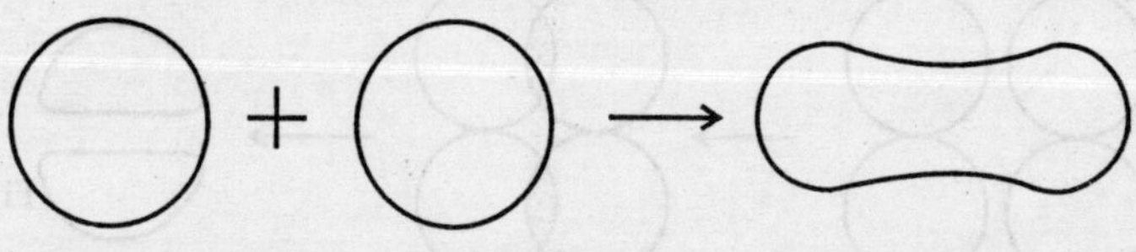

Fig. 3—Formation of σ MO in a hydrogen molecule by s-s coaxial overlap.

P-P overlap: When two P– orbitals overlap coaxially say P_x of both atom overlap coaxially then σ– molecular orbitals are formed. Example– Nitrgen atom has three electrons in valence bond i.e. in P- orbitals perpendicular to each other when the two nitrogen come so close that overlapping of orbitals take place resulting in formation of σ- bond. (Fig. 4)

Fig. 4—Formation of σ MO by coaxial overlap of p orbitals. A-isolated head-on p orbitals; B-coaxial overlap; C-the σ MO.

Formation of pi (π) molecularorbital

When two parallel p atomic orbitals of the two σ bonded

atoms overlap sideways (Fig. 5), the resulting MO is termed as Pi (π) molecular orbital. The electron pair residing in this πMO is called π electron pair, and the bond so formed is called bond. On the other hand 2p-orbitals of each electron overlap sideways since they are parallel resulting πMO that contains electronpair as shown in Ifiy 6B This plane is the modal plane in which π electrons are not found. The π electron pair encompassing the two nuclei is said to be in the localised state.

The *sideways overlapping* of p orbitals, being poorer than the more intimate co-axial overlapping in the formation of σ orbital, the π electrons have greater freedom of movement and so the π MOs are less stable than σ MOs. The π MO have a tendency to switch over to σ by combining with other atoms. This accounts for the unsaturation of olefins and addition reactions exhibited by them.

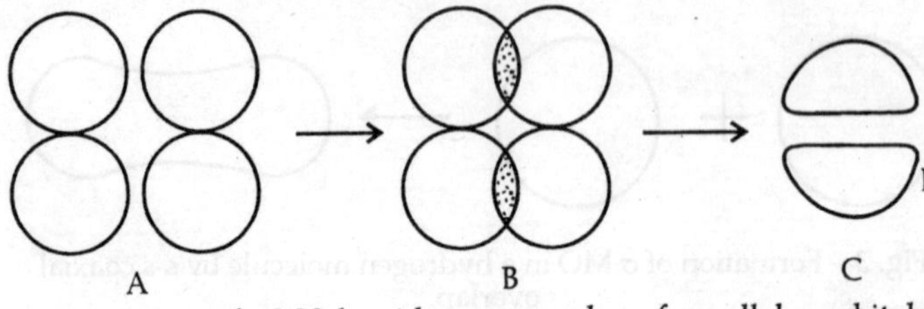

Fig. 5—Formation of π MO by sideways overlap of parallel p-orbitals; A-the parallel p orbitals before overlap; B-sideways overlap; C-The π MO.

Because of their greater mobility, the π electrons and so the π MOs can, under certain conditions, encompass more than two nuclei. This state of electron of MO is known as delocalised state and results a phenomenon of reasonance or mesomerism discussed later in chapter 3.

Oxygen molecule—Orbital Representation

Ground state electronic configuration of oxygen atom is $1s^2\ 2s^2\ 2p^2x\ 2p^1{}_y\ 2p^1{}_z$. The single electron in 2pg and 2pz orbitals take part in bond formation and the paired electrons in is 2s, 2px, are core or non-bonding electrons.The 2p orbitals of the two different oxygen atoms overlap coaxially (Fig. 6A) to form a σ MO accommodating the σ electron pair. On the other hand $2p_y$-orbitals of each electrons overlap sideways since they are parallel resulting π MO that contains π electron pair as shown in (fig 6B).

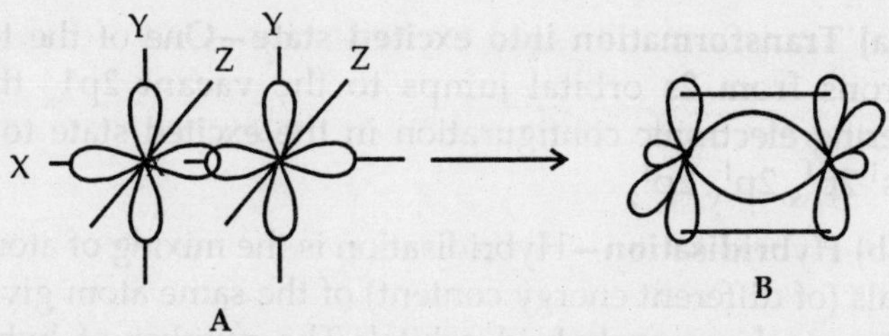

Fig. 6—Molecular orbitals in the oxygen molecule—A. coaxial overlap of p_x orbitals; B. sideways overlap of p_y- p_y and p_2 - p_2 orbitals.

Nitrogen molecule—Orbital Representation

Ground state electronic configuration of nitrogen atom is $1s^2\ 2s^2\ 2p^1{}_x\ 2p^1{}_y\ 2p^1{}_z$ Since , $2p_x$, $2p_y$ and $2p_z$ orbitals contains unpaired electrons so these form bonds, Whereas those present in Is and 2s are non-bonding or core electrons. The $2p_x$ of the two nitrogen atoms overlap coaxially (Fig. 7A) to form σ-MO. The $2p_y$ and $2p_z$ orbitals, perpendicular to each other, are left on each of the nitrogen atoms. The $2p_x$ and $2p_z$ orbitals of each nitrogen atom being parallel overlap sideways resulting in two π MOs perpendicular to each other and each one containing one π electron pair i.e. total of 4-electrons. One σ and two π MOs together represent a triple bond.

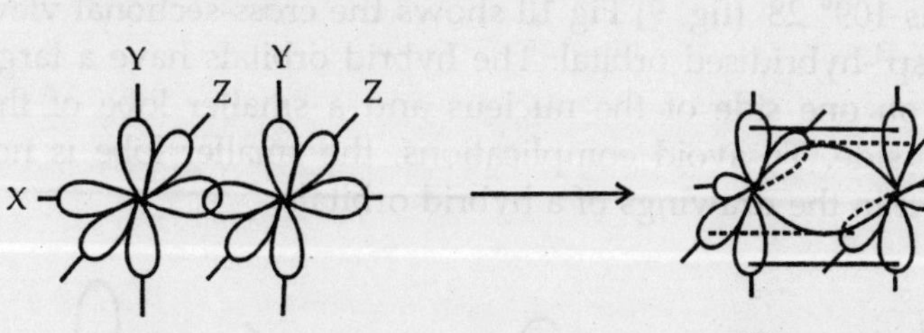

Fig. 7—Molecular orbitals in the nitrogen molecule: A—coaxial overlap of p_x orbitals; B—sideways overlap of p_y-p_y and p_z-p_z orbitals.

Orbital representation of simple organic molecules

Ground state electronic configuration of carbon atom is $1s^2\ 2s^2\ 2p^1{}_x\ 2p_y{}^1$. A carbon atom is not capable of bond formation is its ground-state, so to form g bonds corbon atom has to undergo the following two successive changes:

(a) Transformation into excited state—One of the two electrons from 2s orbital jumps to the vacant $2p1_{\pi}$ thus rendering electronic configuration in the excited state to be $1s^2\, 2s^1\, 2p^1_x\, 2p^1_y\, 2p^1_z$.

(b) Hybridisation—Hybridisation is the mixing of atomic orbitals (of different energy content) of the same atom giving rise to equal-energy hybrid orbitals.The number of hybrid orbitals that results is equal to the number of participating atomic orbitals. The hybrid orbitals are still atomic orbitals and remain oriented around a single nucleus. Bond-formation occurs when the hybrid orbitals overlap the atomic orbitals, hybrid or not, of other atoms. Hybridisation in carbon atom may occur in three different ways depending on the number of atoms it combines with:

(i) sp^3 hybridization—In sp^3 hybridization a carbon atombonds with four other atoms which involves 2s and all the three 2p orbitals in excited state (Fig 8) It gives rise to four new hybrid orbitals called sp^3-hybrid orbitals—each possessing one electron. These new orbitals are equivalent in energy content and are directed towards the vertices of a regular tetrahedron and thus angular distance between any two is 109° 28' (fig. 9) Fig 10 shows the cross-sectional view of a sp^3-hybridised orbital. The hybrid orbitals have a large lobe on one side of the nucleus and a smaller lobe of the other side. To avoid complications, the smaller lobe is not drawn in the drawings of a hybrid orbitals.

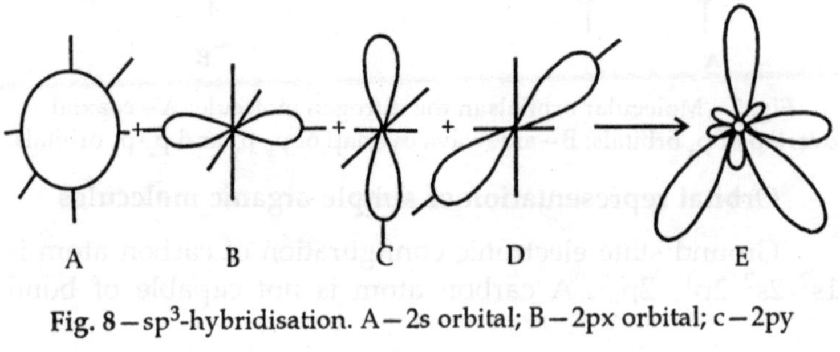

Fig. 8—sp^3-hybridisation. A—2s orbital; B—2px orbital; c—2py

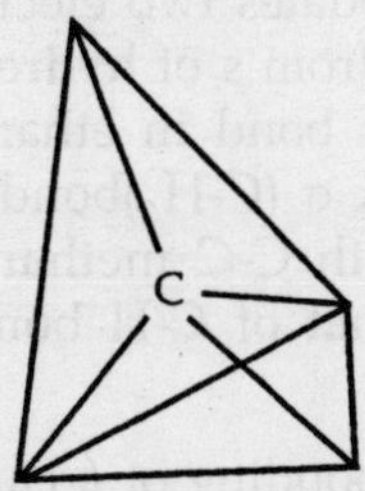

Fig.9 – Four sp3-hybrid orbitals pointing towards the vertices of a regular tetrahedron.

Fig. 10 – Cross section of a sp^3-hybrid orbital.

orbital; D – 2pz orbital; E – Four sp^3-hybrid orbitals.

Bond-formation by sp^3-hybrid orbitals of carbon is illustrated in the following examples –

Methane (CH_4)As shown in fig.11 ,the four tetrahedrally clirected sp^3 hybrid orbitals of carbon overlap coaxially with is orbitals of four Hydrogen (H)-atoms to form four σ(C-H)

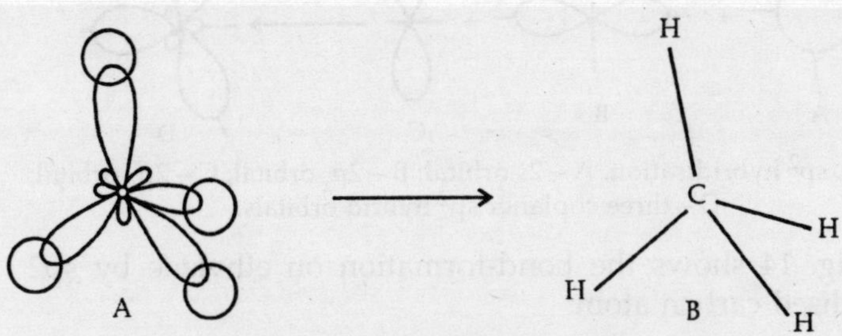

Fig. 11. Bond-formation in methane molecule: A. sp^3 – s overlap B. representation of the tetrahedral structure.

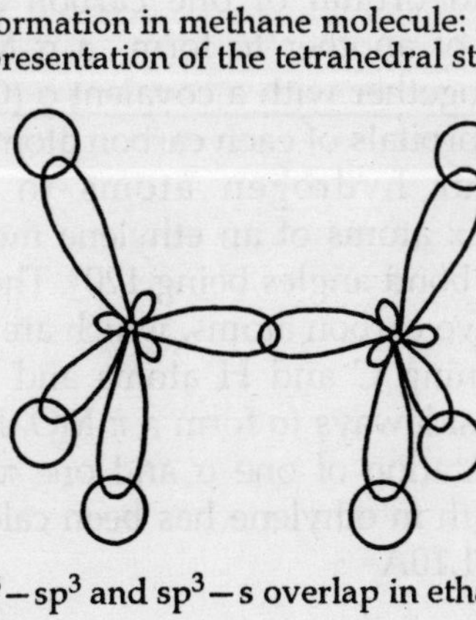

Fig. 12. sp^3 – sp^3 and sp^3 – s overlap in ethane molecule.

molecular orbitals. Each σ MO accommodates two electrons – one from sp^3 of carbon and the other from s of hydrogen. Ethiane(H_3C-CH_3): The carbon-carbon σ bond in ethane is formed by $sp^3 - sp^3$ overlap, and the six σ (C-H)bonds by $sp^3 - s$, overlap (Fig.12) The bond length C-C methane is approximatly found to be 1.54Å and that of C-H bond is 1.11Å

(ii) sp^2-hybridisation– It involves bonding of a carbon atom with only three other atoms. In this to hybridisation of 2s and two 2p (*i.e.* $2p_x$ and $2p_y$) orbitals take place (Fig.13). It gives rise to three coplanar sp^2- hybrid orbitals of equal energy content. These are triagonal *i.e.* they form an angle of 120° with each other. Each of these new hybrid orbitals has one unpaired electron. The unhybridised $2p_x$ orbital with its single electron remains at right angle to the plane of three sp^2 hybrid orbitals (not shown in the figure).

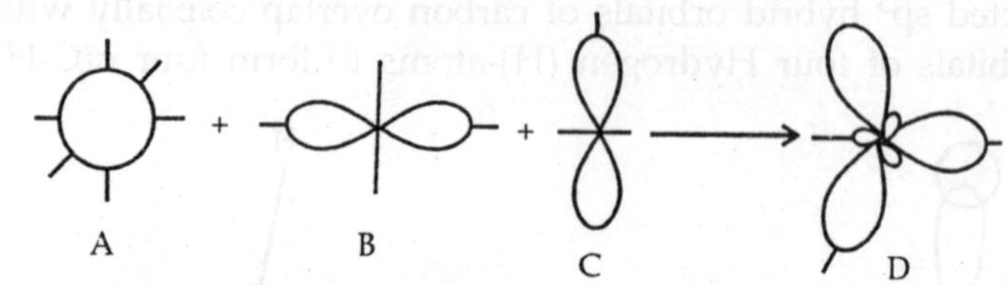

Fig. 13. sp^2 hybridisation. A – 2s orbital; B – $2p_x$ orbital; C – $2p_y$ orbital; D – three coplaner sp^2 hybrid orbitals.

Fig. 14 shows the bond-formation on ethylene by sp2 hybridised carbon atom.

One sp^2 hybrid orbital of one carbon atom overlaps coaxially with that of another to form a σ MO linking the two carbon atoms together with a covalent σ (C-C)bond. The left -out sp^2 hybrid orbitals of each carbon atoms overlap with 1s orbitals of four hydrogen atoms to form four σ (C-H) bonds. The six atoms of an ethylene molecule thus lie on the same plane , bond angles being 120°. The unhybridised $2p_x$ orbitals of the two carbon atoms, which are at right angles to the plane containing C and H atoms and are parallel to each other, overlap sideways to form a π MO. Hence C = C is a symbolic representation of one σ and one π MO together. The C=C bond length in ethylene has been calculated as 1.34 A and of C-H to be 1.10A

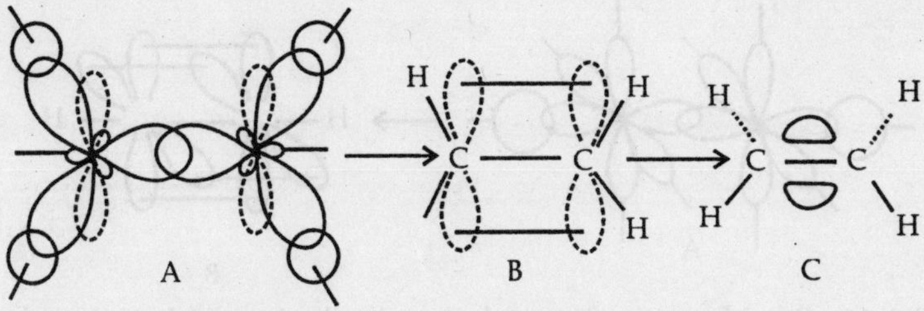

Fig. 14—Bond formation in ethylene molecule. A—coaxial overlap of one sp^2-sp^2 and four sp^2-s orbitals; B-sideways overlap of parallel px orbitals; C—π MO in ethylene molecule

(iii) sp hybridisation—Carbon atom, bonding with only two other atoms undertakes hybridisation of the 2s with any one of the three 2p-orbitals, say $2p_x$, to form two collinear **sp-hybrid** orbitals (Fig.15) of equal energy content. There hybrid orbitals are diagonally dircted (180° apart) and each of them has one unpaired electron.

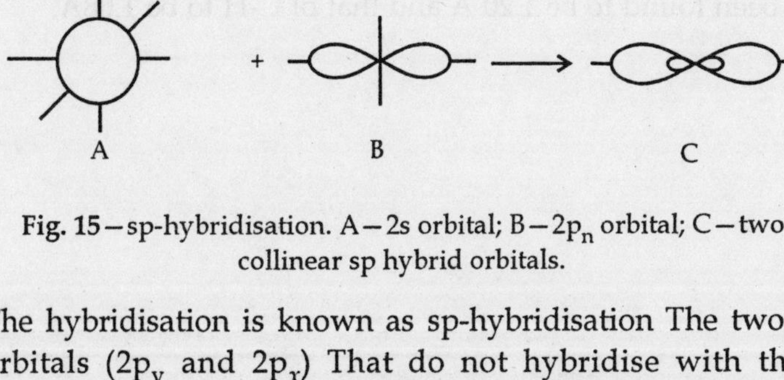

Fig. 15—sp-hybridisation. A—2s orbital; B—$2p_n$ orbital; C—two collinear sp hybrid orbitals.

The hybridisation is known as sp-hybridisation The two p-orbitals ($2p_y$ and $2p_x$) That do not hybridise with their unpaired electrön (one in each) are at right angles to each other and also to the line of sp-hybrid orbitals (Not shown in fig)

Fig.16 shows the bond -formation by sp-hybridised carbon atoms in a cetylene(C_2H_2).

One sp-hybrid orbital of a carbon atom overlaps coaxially with that of another to form a σ MO, linking the two carbon atoms together with a covalent σ (C-C) bond. The remaining

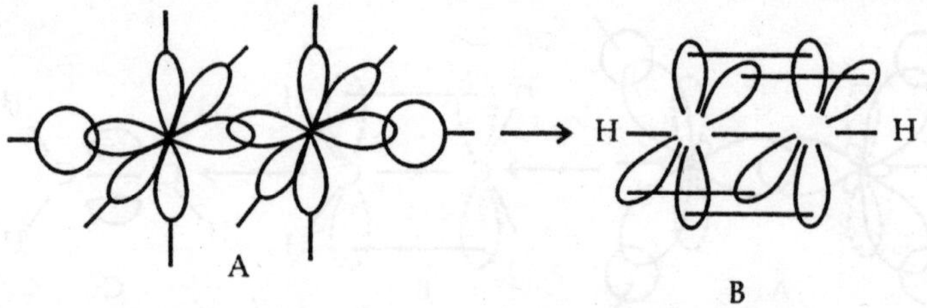

Fig. 16 — Bond formation in acetylene molecule. A-coaxial overlap of sp-sp and sp-s; B-sideways overlap of p_y-p_y and p_z-p_z orbitals.

one sp-hybrid orbital of each of the two carbon atoms overlaps with 1s orbitals of two hydrogen atoms to form two σ (C-H) bonds. The two $2p_x$ and $2p_z$ orbitals which are normal to each other are, thus left on each of the C atoms. Tlie $2p_y$ orbitals of each C atom, being parallel, mutually overlap sideways, forming one π MO. Similar overlapping occurs between the two mutually parallel $2p_z$ orbitals to form another π MO. These two π MOs so formed are perpendicular to each other. One σ and two MOs collectively represents a triple bond *i.e.* C≡C. The bond length of C≡C in acetylene has been found to be 1.20 A and that of C-H to be 1.08A.

Chapter 3
RESONANCE

π-Molecular Orbitals Delocalisation

In chapter 2, we have studied that molecular orbitals in which the bonding electron-pair is associated with only two atoms. Such molecular orbitals are said to be localised MOs.

When more than two parallel p-orbital are present on adjacent and coplanner σ–bonded atoms, the p-orbital in the middle may undergo sideways overlapping in the middle may undergo sideways overlapping with the adjacent p-orbitals on either side to form extended π molecular orbital in which electrons encompass more than two atoms. Such molecular orbitals are said to be delocalised π MOs.

The phenomenon of delocalisation of π molecular orbitals has been explained below taking the example butadiene (C_4H_6)

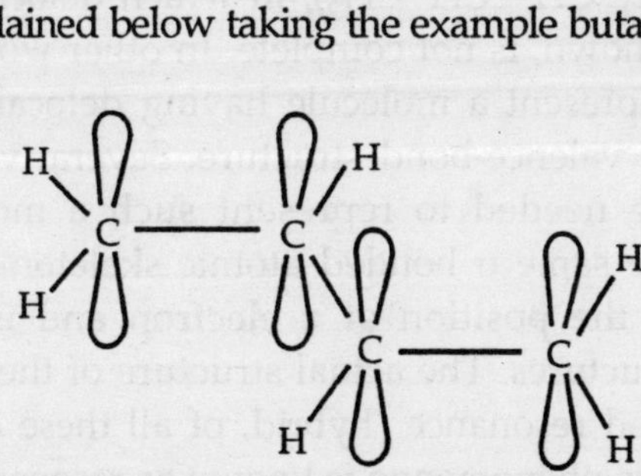

Fig. 17 – σ bonded C and H aloms in the same plane, with perpendicular and mutually parallel p-orbitals.

In butadiene ($\overset{1}{C}H_2 = \overset{2}{C}H - \overset{3}{C}H = \overset{4}{C}H_2$) each carbon atom is sp^2 hybridised having a p-orbital perpendicular to three coplanar sp^2 hybrid orbitals. Sp2 hybrid are used to form σ (C-C and C-H) bonds leaving four parallel p-orbitals with one on each carbon atom Fig (17).

Fig 18 show that the parallel p-orbitals on carbons C_2 and C_3 may overlap mutually as well as with the p-orbitals on C_1 and C_4 respectively to form delocalised π MOs which are extended over all the four carbon nuclei (Fig.18) Since no orbital can accomodate more than two electrons and there are two pairs of π electrons in butadiene, the number of delocalised π MOs is also two in this case.

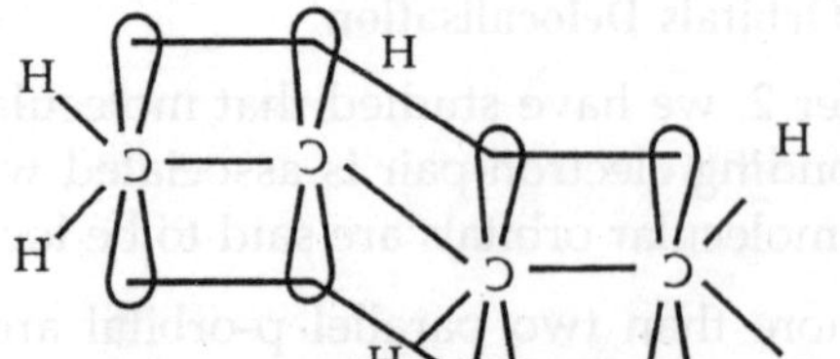

Fig. 18 – Delocalised π MOs in butadiene

Hence, delocalisation of π MOs leads to the C-C bonds in butadiene acquire a character between a single bond and a double. Representation of butadiene by valence-bond structure (CH_2=CH–CH=CH_2), in which delocalised σ and π bonds are shown, is not complete. In other words, it is not possible to represent a molecule having delocalised π MOs by any single valence-bond structure. Several valence-bond structures are needed to represent such a molecule. The structure have same σ bonded atomic skeleton but there is difference in the position of π electron and is known as resonating structures. The actual structure of the molecule is a hybrid, called resonance hybrid, of all these contributing structures. This phenomenon is known as resonance. Double-headed arrows are drawn in between two resonating structures to represent the hybrid. Resonating structures of butadiene are given below:

$$CH_2{=}CH-CH{=}CH_2 \leftrightarrow \overset{-}{\ddot{C}}H_2-CH{=}CH-\overset{+}{C}H_2$$
$$\leftrightarrow \overset{+}{C}H_2-CH{=}CH-\overset{-}{\ddot{C}}H_2$$

Evidently, resonance is another way of representing delocalisation of π MOs.The concept of resonance is arbitrary but it provides a great help in explaining or predicting the properties of molecules.

Delocalisation of π MOs or the phenomenon of resonance is found in those molecules in which a minimum of three p-orbitals are parallel to each other and perpendicular to the plane of the σ bonded adjacent atoms. Such a system of p-orbitals exists in the valence-bond structures where—

(A) *Alternate single and double bonds* **(conjugated double bonds)** *are present between coplaner atoms,* such as in butadiene1, 3; pentadiene – 1, 3; acrolein (CH_2=CH-CHO), benzene, benzaldehyde, benzoic acid and nitrobenzene etc.

(B) *An unshared pair of p-electrons is present on an atom which is adjacent and coplanar to π bonded atoms,* such as in vinyl chloride (CH_2=CH – Cl), carboxylate anion, carbonate anion, aniline, phenol and chlorobenzene etc.

(C) *A positive-charged sp^2 hybridised atom is adjacent and coplanar to π bonded atoms,* ,such as in allyl carbonium ion and benzyl carbonium ion etc.

To understand easily and write resonating structures, a curved arrow (↷) is used to show the formal shift of π or p-electron pairs. These arrows are only formal and have no physical significance. Resonating structures of some molecules are given below to illustrate resonance. Take care of the fact that the atoms and bonds do not move at all and complete structures must have the same number of electron pair.

Case I : 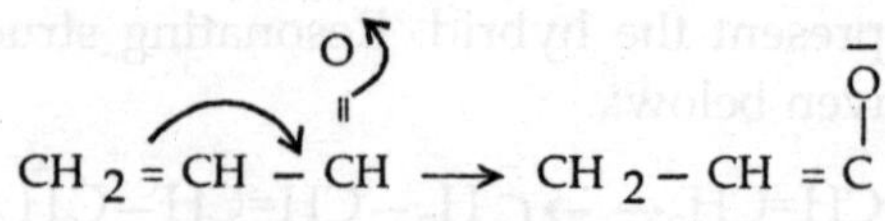

(Acrolein)

Case II : 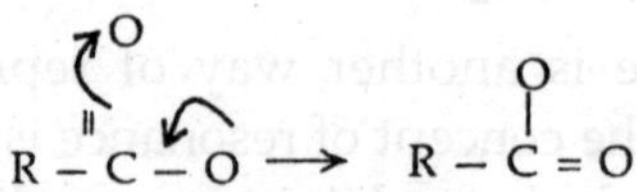

(Carboxylate anion)

Case II : 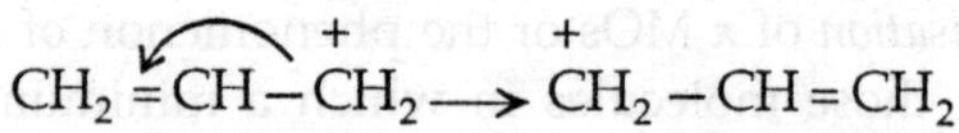

(Allyl carbonium ion)

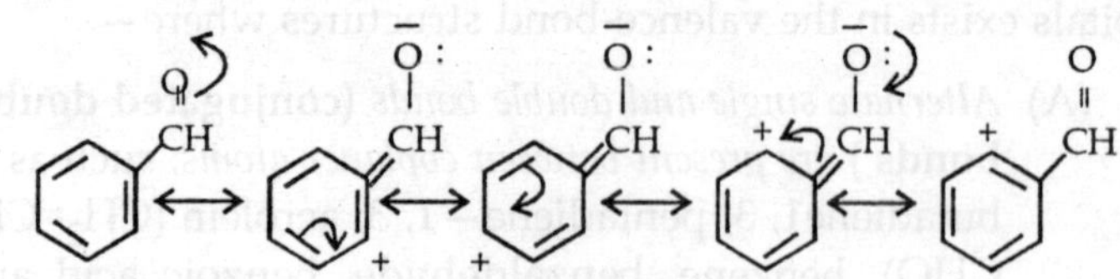

(Benzaldehyde)

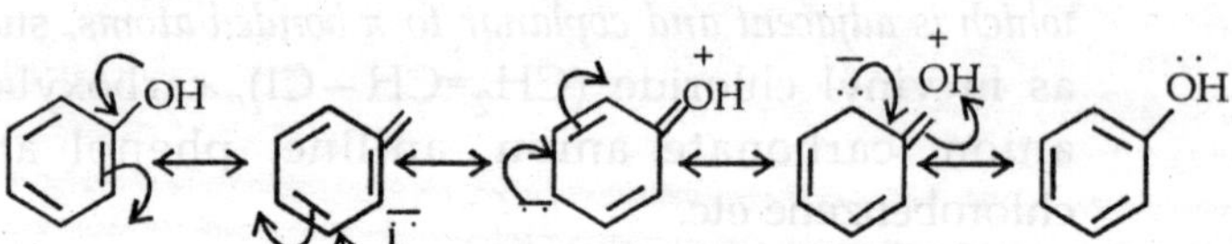

(Phenol)

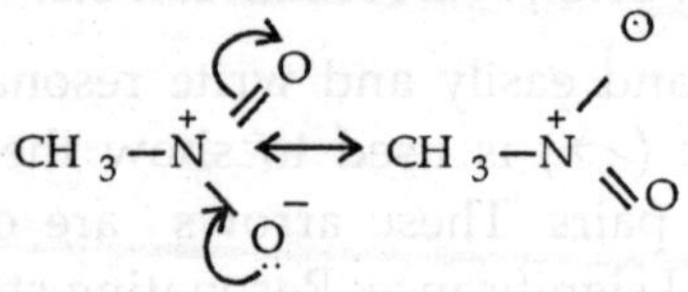

(Nitromethane)

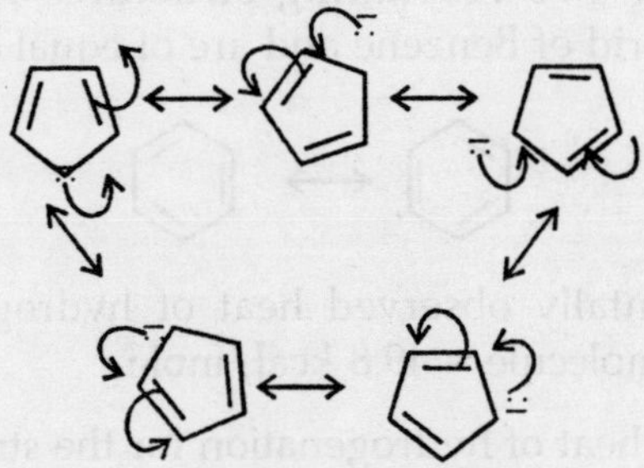

(cyclobutadienyl anion)

Results of resonance

As a result of delocalisation of MOs (or resonance) in a molecule, following peculiarities are observed:

(a) *Stabilisation* – Electrons in polycentric delocalisedπ MOs have a greater freedom of movement than those in dicentric localised ones. Hence π electrons attachment is extended to a greater number of nuclei and the resulting delocalised π MOs become more stabilised. In other words, the real molecule (or the resonance hybrid) in which π MOs are in a delocalised state are more stabilised than one represented by any resonating structures which exhibits localised π electrons. This *stabilisation due to delocalisation* is called **resonance stabilisation**.

The difference between experimentally calculated energy of a molecule and the energy of most stable of its resonating structures is called resonance energy of delocalisation. Higher the resonance energy, greater is the stabilisation of the system.

Resonance energy of a molecule may be determined by calculating the difference of the observed heat of hydrogenation of the actual molecule and the calculated heat of hydrogenation corresponding to the number of double bonds shown in most stable of its resonating structues.Resonace energy calculation of benzene molecule is described below.

Following two resonating structures represents the resonance hybrid of Benzene and are of equal energy:

Experimentally observed heat of hydrogenation for a benzene molecule = 49.8 kcal/mole

Predicted heat of hydrogenation for the structure having three double bonds i.e. cyclohexatriene-1, 3, 5 = 3 × 28.6* = 85.8 kcal/mole

∴ Resonance energy of benzene = 85.8 − 49.8 = 36.0 kcal/mole

Table 1 shows the resonance energies of some common molecule.

On the basis of following points the extent of stabilisation of a delocalised system is determined.

(i) The resonating structures of higher energy, like ions or free radicals are poor contributors.

TABLE 1. RESONANCE ENERGIES (Kcal/mole)OF SOME COMMON MOLECULES

cyclohexadiene-1, 3	1.8	benzene	26
cyclopentadiene-1, 3	2.9	acetophenone	36
butadiene-1, 3	35	azulene	37
acetic acid	14	styrene	38
acetamide	17	aniline	40
ethyl acetate	18	pyridine	43
furan	22	indole	49
tropolone	22	naphthalene	61
pyrrole	24.5	quinoline	69
carbon dioxide	27	diphenyl	71.5
hiophene	28	trans-stilbene	78
acetic anhydride	30	anthracene	84
phenyl acetylene	35	phenanthrene	92
toluene	35	carbon monoxide	105
phenol	36	1,3,5-triphenyl benzene	150

(ii) When the contributing structures are all equivalant,as in benzene or RCOO, the resonance energy is greater and so, more is the stabilisation.

(iii) More is the number of resonating structures of equal energy more is the stabilisation.

(iv) Systems having charge-separation in their resonating structures are less stabilised than those having no charge separation. RCOO, for example, is more stabilised than RCOOH.

(v) Resonating structures having negative charge on same atom makes the system more stabilised than those with negative charge on dissimilar atoms. For example: RCOOH is more stabilised than C_6H_5-O.

(b) *Dual nature of bonds.*—Because of greater freedom of movement of delocalised π electron, the bonds between the encompassed atoms are neither purely single nor purely double but are hybrid of the two. The data of bond length measurement show and supports the previous attachment. In benzene, for example, all the six carbon-carbon bonds are of the same length *i.e.* 1.39A, intermediate between 1.48A for a single and 1.34 A for a double bond. The length of nitrogen-oxygen bonds in nitromethane is found equal *i.e.*1.21 A-but the normal length of nitrogen-single bond 1.36A and of nitrogen-oxygen double bonds 1.18A.

Aromaticity

Benzene shows unusually high stability*(resonance energy=36kcal/mole) and inspite of having high unsaturation it undergoes electrophilic substitution rather than addition. Same characteristics are shown by the derivatives of benzene.This feature is referred to as aromaticity and the compounds possesing this character are said to be aromatic. These characteristics are attributed to resonance stabilisation(or delocalisation of π MOs. Fig. 19).

Sextet (6 π electrons) is only one of the many π electron systems which confer aromaticity to the molecule. According

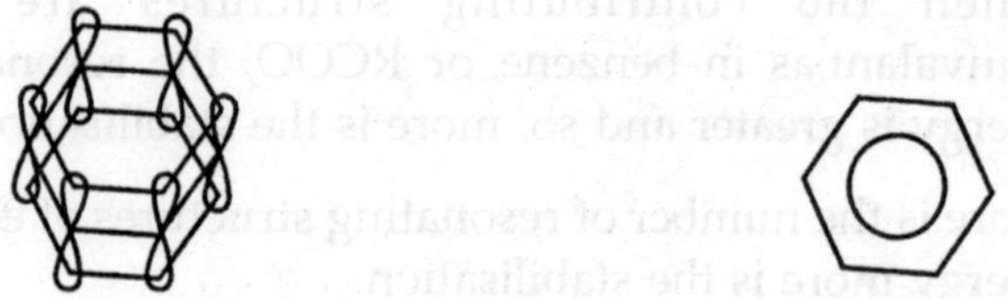

A B

Fig. 19.—A. Sideways overlap of six parallel p-orbitals, standing on six coplanar and adjacent σ—bonded carbon atoms, on either side to form three delocalised π MOs, B. Graphical representation of circular delocalised π MOs.

to Huckel, all coplanar cyclic systems with conjugated double bonds would be aromatic if they possess (4n+2) π electrons, where n = 0 1, 2, 3... This is known as **Huckel's rule** for aromaticity. Hence the rings with 2, 6, 10, and 14 π electrons are aromatic while those with 4, 8, 12 and 16 etc. are not.

Benzene,naphthalene and pyridine, 6π, 10π and 14π electrons respectively according to the (4n+2) π electron rule which is favourable for resonance stabilisation and hence are aromatic.

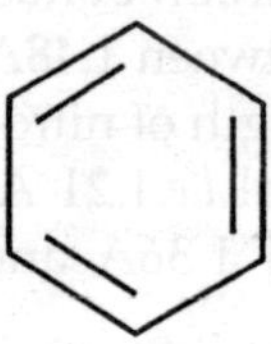

benzene
(6 π electrons)

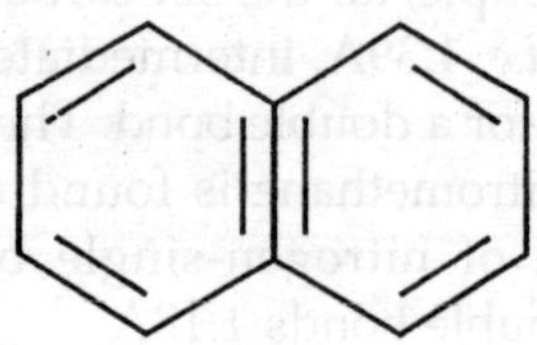

naphthalene
(10 π electrons)

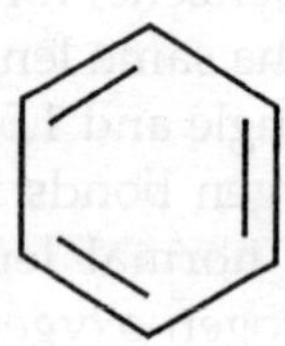

pyridine
(6 π electrons)

Cyclobutadiene and cyclooctatetraene have neither all the ring atoms in one plane (*due to internal strain*), nor (4n+2)π electrons. The above compounds do not follow (4n+2)π electron rule so not stablised by delocalisation of π MOs and are not aromatic.

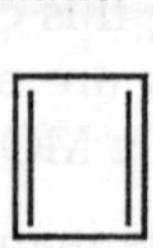

cyclobetadieae—1,3
(4π electrons)

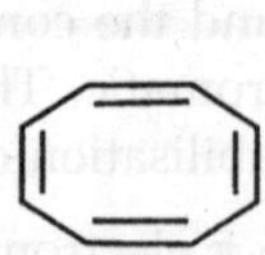

Mooctaterraene—1, 3, 5, 7
(8π electrons)

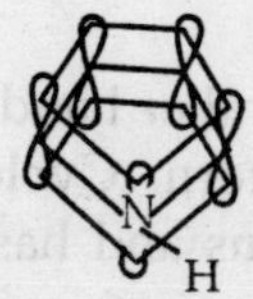

In pyrrole, four carbon atoms and one nitrogen make a coplanar σ bonded ring. Each of the four carbon atom contains one electrons in its one p-orbital and nitrogen has a p-orbital with an electron pair. All these five parallel p-orbitals overlap sideways to form three delocalised π MOs accommodating 6 π electron (Fig. 20). This 4n+2 pattern imparts aromatic character to pyrrole.

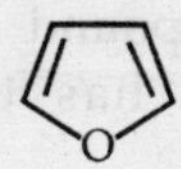

Thiophene

Cyclopentadiene

Other similar five-membered heterocyclic compounds such as furan and thiophene acquire aromaticity because of the sextet of electrons made of two pairs of π electrons from the two double bonds and one pair from the hetero atom.

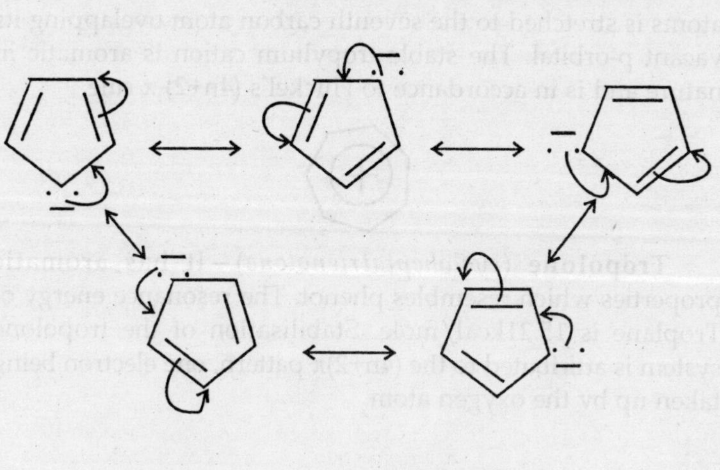

(Resonance – stabilised Cyclopentadienyle ion)
Non-benzenoid aromatic compounds

Cyclopendiene, on the other hand is non-aromatic as it does not possess 4n+2 pattern. Cyclopentadienyl anion does contain (4n+2)p electrons and has the stabilisation required for aromaticity.

Fulvenes—6, 6-dimethyl fulvene (Resonance energy =11.9 kcal/mole) has a moderate aromatic character. It is the exocyclic double bond that makes available to the five-membered ring a sextet of electroons.

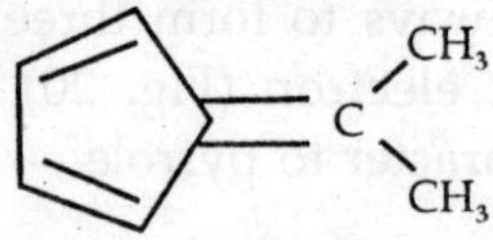

Ferrocene (*dicyclopentadienyl iron*)—The compound has more aromatic character than benzene. It has two cyclopentadienyl ions linked by iron.

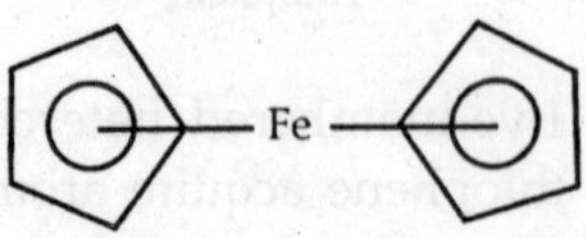

Tropylium cation (*cycloheptatrienyl cation*) In tropylium cation, the delocalised 6 π electron system over six carbon atoms is stretched to the seventh carbon atom ovelapping its vacant p-orbital. The stable tropylium cation is aromatic in nature and is in accordance to Huckel's (4n+2) π rule.

Tropolone (*cycloheptatrienolone*)—It has aromatic properties which resembles phenot. The resonance energy of Troplane is 15.21kcal/mole. Stabilisation of the tropolone system is attiributed to the (4n+2)π pattern, one electron being taken up by the oxygen atom.

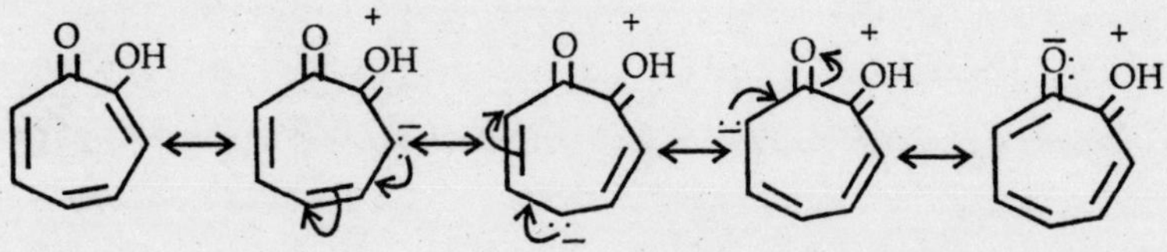

Azulene—Azulene is a crystalline solid blue in colour and resembles in odour with its isomer naphthalene. Its resonance energy is 49, kcal/mole Its an aromatic compound which has 10 π electrons and is in accordance with Huckel's rule. The structures contributing to its resonance hybrid are:

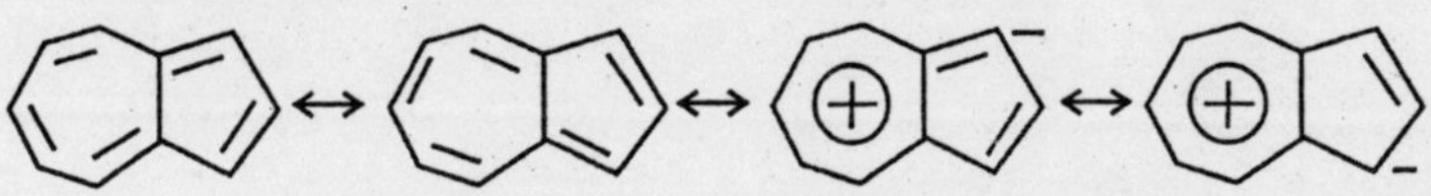

Chapter 4

FACTORS AFFECTING BOND-REACTIVITY

Electronegativity and Polarisation of bond.

The extent of attraction of outer-shell electrons by the nucleus is called electronegativity of the atom. In a periodic table the electronegativity increases from left to right with increasing nuclear charge in any horizontal row. In any vertical group, electronegativity decreases from top to bottom as the outer shell becomes distant from the nucleus. The electronegativity values for some common elements are as follows:

H	P	C	S	I	Br	N	CI	O	F
2.1	2.1	2.5	2.5	2.5	2.8	3.0	3.0	3.5	4.0

When two atoms of differing electronegativity are linked by a covalent bond, the bonding electron pair is attracted more towards the atom of higher electronegativity i.e. the charge distribution between the two atoms is uneven. The atom of higher electronegativity has a fractional negative charge (δ^{-1}) and that of lower electronegativity acquires fractional positive charge (δ^{+}). The bond in such a state is said to be polarised *i.e.* it acquires a partial ionic character. If the difference in electronegativity between bonded atoms, is more, its polarity is more. The bond polarisation is depicted by an arrow along the bonding line. For example, in methyl chloride, the C*l* atom qcquires a δ^- charge and carbon atom of methyl acquires a δ^+charge:

$$H_2\overset{\delta_+}{C} - \rightarrow - \overset{\delta_-}{Cl}$$

Inductive effect

Two dissimilar atoms joined by a covalent bond develop opposite fractional charges (δ^+ and δ^-)due to polarisation of bond. If such a fractionally charged atom is linked with other atoms,smaller like charges are induced along the chain, viz.,

$$\begin{array}{c}
| \;\; \delta'''_+ \quad | \;\; \delta''_+ \quad | \;\; \delta'_+ \quad \delta^- \\
- C \rightarrow - C \rightarrow - C \rightarrow Cl \\
| \qquad\quad | \qquad\quad |
\end{array}$$

(Where $\delta'_+ > \delta''_+ > \delta'''_+ \qquad \delta'_+ + \delta''_+ - \delta'''_+ = \delta_+$)

This phenomenon of inducing like charges over the adjacent atoms in a chain is called inductive effect or, I effect. The I-effect is a permanent effect in the ground state of a molecule and it operates through σ-bonds. The effect diminishes with increase in distance from the source (in the example cited above, C*l* is the source). If the electron density is pulled towards the source, the effect is said to be$-$I effect of the source and if pushed away from the source, it is + I.

A few atoms/groups are listed below in the decreasing order of their +I or $-$I effects:

+ I effect

$COO > -C(CH_3)_3 > -CH(CH_3)_2$
$> -CH_2CH_3 > -CH_3$

$-$ I effect

$NR_3 > -NO_2 > -CN > -CO- > -F$
$> -Cl > -Br > -OR > -CH=CH_2 > -C\equiv CH$
$> -C_6H_5$

Electromeric effect

It is a temporary effect by π electrons and operates instantaneously on requirement of the attacking reagent. It involves complete transfer of a π electron -pair to one of the two π-bonded atoms. Graphically represented by a curved arrow starting from the original position of the π electron-pair and terminates at either of the π bonded atoms to which this electron pair migrates. The atom that gets the control of the π electrons gains a negative charge and the other atom which has lost its share in the electron-pair becomes positively-charged.

The direction of the displacement of π electrons by electromeric effect is dependent on the nature of certain atoms or groups attached to the π bonded system. If the displacement is towards the group, it is denoted by -E; and if away from the group, it is denoted by +E., A few examples are given below for illustration—

(i) In propene, due to +I effect of methyl group the electromeric displacement of π electron-pair at the requirement of the reagent is away from the methyl group, viz.,

$$CH_2 \to CH \overset{\curvearrowright}{=} CH_2 + \overset{+}{A}\overset{-}{B} \to CH_2 - \overset{+}{C}H - \underset{\underset{A}{|}}{C}H_2 + \overset{-}{B}$$

Methyl group exerts a+E effect on electrons in this case.

(ii) In allyl chloride, due to −I effect of chlorine, the electromeric displacement of π electron-pair is towards chlorine, viz.,

$$Cl \to CH_2 - CH \overset{\curvearrowleft}{=} CH_2 + \overset{+}{A}\overset{-}{B} \to Cl - CH_2 - \underset{\underset{A}{|}}{C}H - \overset{+}{C}H_2 + \overset{-}{B}$$

Chlorine exerts a−E effect on π electron in this case.

(iii) In phenol, the unshared pair of electrons on oxygen of OH, being in conjugation with the conjugated double bonds of the ring, is delocalised towards the ring. At the requirement of the electrophilic reagent*, an electromeric displacement of delocalised π electron, therefore, takes place towards the ring atoms (Fig. 21). OH

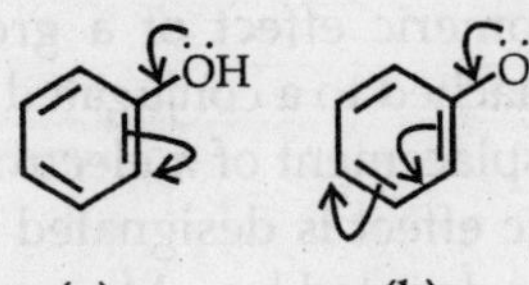

Fig. 21 – Electromeric displacement of π electrons in phenol (a) to *o*-position and (b) to *p*-positions.**

group exerts a+E effect on the π electrons of the ring in phenol. In the same fashion NH_2 in aniline and Cl in chlorobenzene exert +E effect on the π electrons of the ring.

(iv) In benzaldehyde, the π electrons of the ring are delocalised towards the π electrons of the conjugated carbonyl group (C=O) due to greater electronegativity of oxygen. At the requirement of the electrophillic reagent, an electromeric displacement of the delocalised π electrons, therefore, takes place towards the carbonyl group (Fig. 22). In benzaldehyde carbonyl group exerts –E effect on the π-electrons.

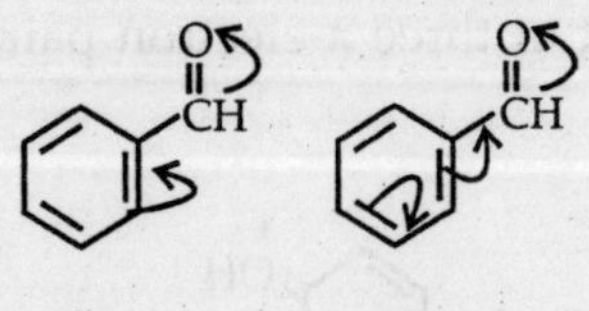

Fig. 22 – Electromeric displacement of p electrons in benzaldehyde: (a) to *o*-position and (b) to *p*-position.

---

* Electron-deficient reagent (see chapter 5)

** Electromeric displacement of p electrons is propagated along the conjugated double bonds..

$\overset{+}{N}O_2$ in nitrobenzene, COOH in benzoic acid and SO_3H in benzene sulphonic acid exert–E effect on the π electrons of the ring similarly.

Mesomeric effect

It is a permanent effect which involves the displacement of π electrons in a molecule exhibiting resonance (or mesomerism+. Mesomeric effect of a group operates only when the group is attached to a conjugated system of multiple bonds. When the displacement of π electrons is *away from the group,* the mesomeric effect is designated by +M, and when *towards the group,* it is denoted by – M.

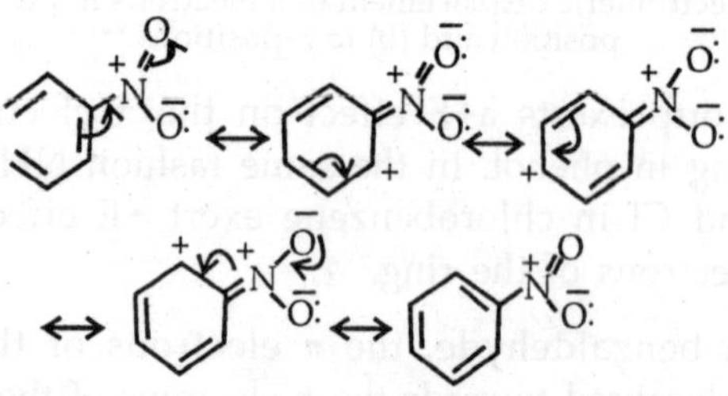

Fig. 23 – Resonating structures of nitrobenzene.

In nitrobenzene, for example, – NO_2 group exerts a – M effect on the conjugated system of double bonds in the ring, because the π electrons are displaced towards the group. In three of the five resonating structures of nitrobenzene (Fig. 23), the ring has a decreased electron density as one out of three π-electron pairs migrate towards NO_2 group.

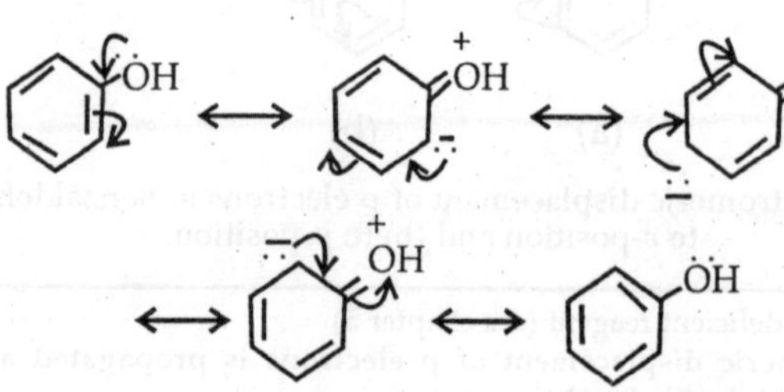

Fig. 24 – Resonating structures of phenol.

On the other hand, −OH group in phenol has a +M effect, because the lone pair of p electrons on oxygen is displaced away from the group on account of conjugation of these electrons with the π bonded system of the ring. As shown in fig 24. out of five three resonating structures of phenol, the ring has increased electron density as a result of transfer of the lone pair of p electrons towards the the ring.

Groups like NO_2, CN, COOH, COOR, $CONH_2$, CHO, COR and SO_3H exert −M effect, while $\bar{O}$, NH_2, NHR, NR_2, NHCOR, OH, OR, OCOR, Br, I and Cl exert +M effect.

Hyperconjugation

If a C−H bond is attached to a double-bonded carbon atom viz, H − C − C = C −, the σ electron-pair of the C−H bond gets delocalised to a certain extent due to extended overlap of the π MO with the adjacent σ MO (Fig. 25) This type of delocalisation is called hyperconjugation or Baker−Nathan effect after the name of its propounders.

Fig. 25−Overlap of s (C−H) MO with p MO in propene.

A valence-bond structure showing localised MOs will not give a true picture of a molecule having hyperconjugation. Hence various structures differing in the position of localised MO,s called hyperconjugative structure, are written and the real molecule structures is the resonance hybrid of all the different structures. Hyperconjugative structures* of propene, for example, are given below:

---

* The proton is not rendered free. Instead, it remains encompassed by delocalised electrons of the system (Fig. 25).

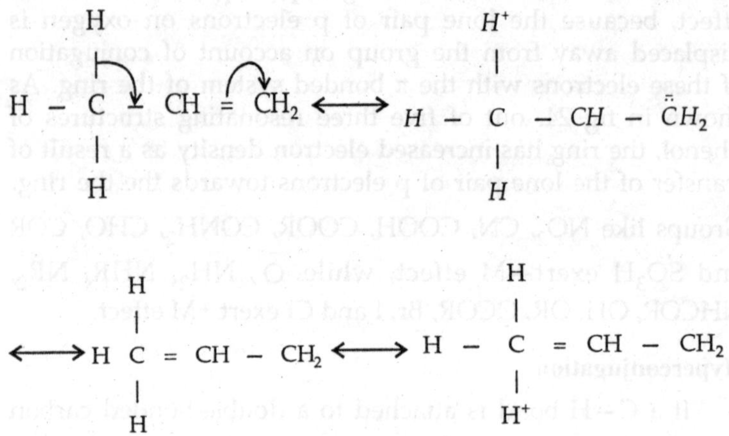

No bonds are shown between C and H in the hyperconjugative structures. Hence, hyperconjugation is also called No-bond resonance.

It may be expected that CH_3-C bond shortens in propane because it has acquired double bond character. Bondlength measurement data confirm it, viz., CH_3-C bond length in propene is found to be 1.50A as compared to 1.54. A for a pure single bond.

Hyperconjugation stabilises a molecule due to resonance. The greater the number of (C-H) bonds attached to a double bonded carbon atom, more the number of hyperconjugative structures and so greater is the stabilisation of the molecule.

Note: – σ (C – H bond is also delocalised due to hyperconjugation if it is attached to

(i) a positively-charged carbon atom, viz.

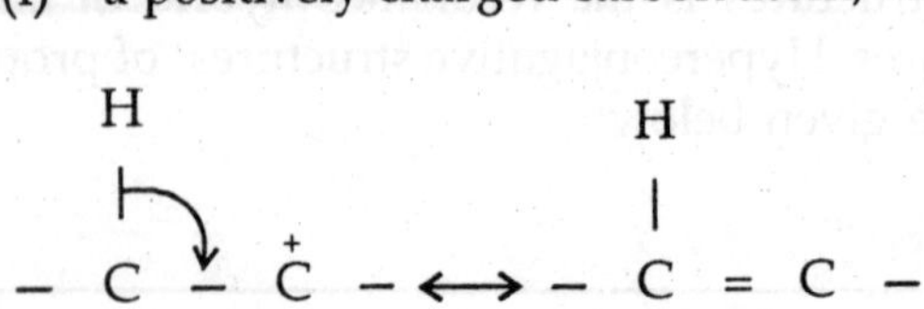

e.g.

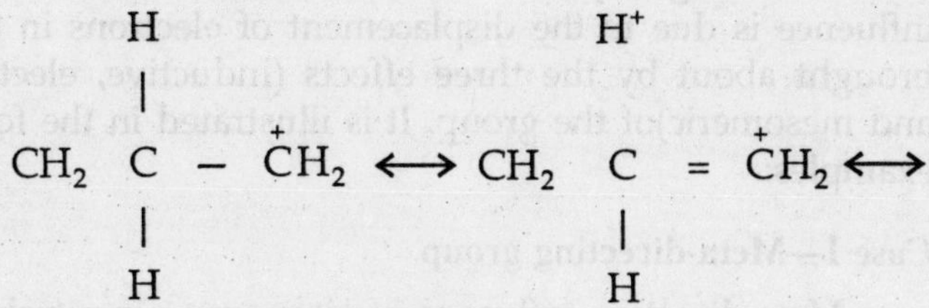

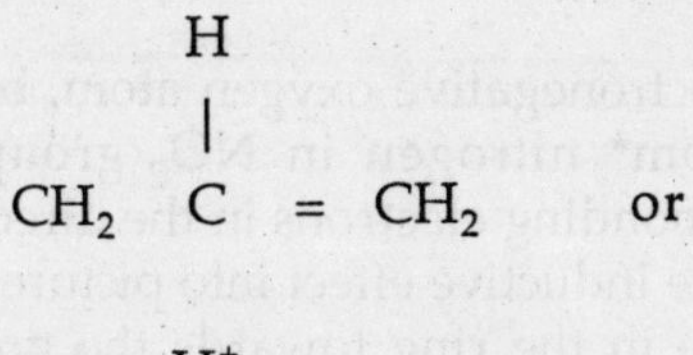

(ii) a carbon atom with an unpaired electron,viz.,

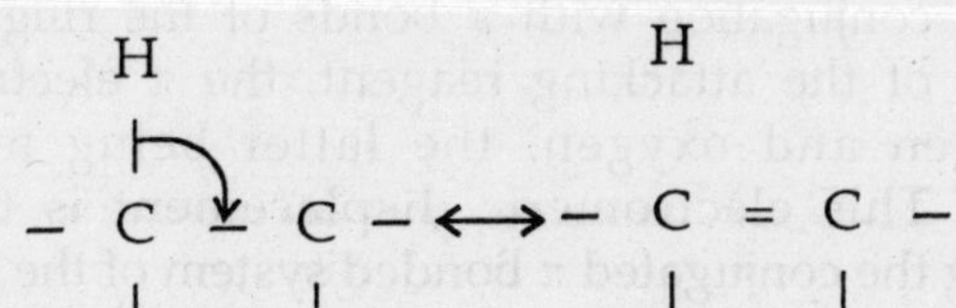

e.g.

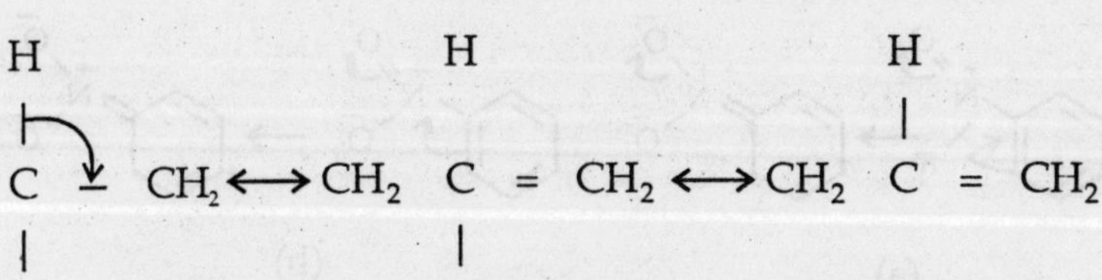

Appplication of the above effects to explain directive influence of groups in aromatic substitutions

A group which is already present in an aromatic ring effects substitution in the ring in two ways—it fixes the position of an incoming group in the ring, as well as well as, renders the process of substitution easier or more difficult. Such influence of group is called its directive influence. This influence is due to the displacement of electrons in the ring brought about by the three effects (inductive, electromeric and mesomeric)of the group. It is illustrated in the following examples:

Case I—Meta-directing group

Meta directing influence of nitro group in nitrobenzene, for example, may be explained on the basis of the three effects as below:

–I effect : A strogly electronegative oxygen atom, being attached with the key-atom* nitrogen in NO_2 group of nitrobenzene, displaces the bonding electrons in the direction towards itself. This brings the inductive effect into picture and the bonding electrons move in the ring towards the group. This –I effect of nitro group decreases the electron density of the ring mainly at ortho positions.

– E effect : The π bond between nitrogen and oxygen of NO_2 group is in conjugation with π bonds of the ring. At the requirement of the attacking reagent, the π electrons between nitrogen and oxygen. the latter being more electronegative. This electromeric displacement is then propagated along the conjugated π bonded system of the ring and brings about the electron-withdrawal from o- and p-positions of the ring to deactivate these positions for electrophilic substitution (Fig. 26).

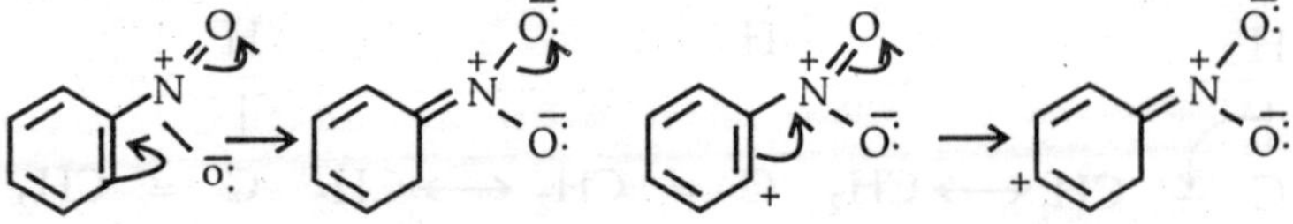

(a) (b)

Fig. 26 – E effect of NO_2 group at (a) *ortho* and (b) para positions in nitrobenzene.

– M effect : The conjugation of π electrons between nitrogen and oxygen of the nitro group with the delocalised

---

* The atom of a group directly linked to the ring is said to be the key atom.

π electrons of the ring brings all the π electrons of the molecule in a state of delocalisation around nitrogen, oxygen and the ring atoms. The structure of nitrobenzene is thus the hybrid of five resonating structures (Fig. 23)

Three of these resonating structures of nitrobenzene have positive charge at o– and p-positions, indicating the electron withdrawal from the ring. This is −M effect of nitro group.

Evidently, therefore, all the three effects of a nitro group act in the same direction *i.e.* towards the group, decreasing electrondensity of the ring mainly at o-and p- positions. Hence the electrophilic substitution is made difficult and oceurs only under extreme conditions at *meta* position because of its relatively higher electron density than at *ortho* and *para* positions.

Note: Similar explanation can be given for meta-directing influence of the groups like—

$$-CN,\ -SO_3H,\ -\overset{\overset{\displaystyle O}{\|}}{C}-H,\ -\overset{\overset{\displaystyle O}{\|}}{C}-R,\ -\overset{\overset{\displaystyle O}{\|}}{C}-OH,\ -\overset{\overset{\displaystyle O}{\|}}{C}-OR\ \textit{etc.}$$

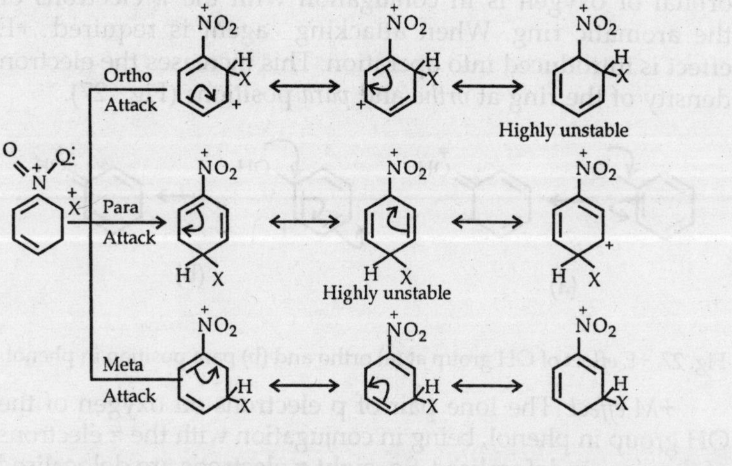

---

* The two adjacent positive centres make the structure highly unstable.

Explanation on the basis of relative stabilisations of the intermediate carbonium ions, formed by the attack of an electrophile X^+ (see mechanism of electrophilic substitutions in chapter 7) at *ortho, para* and *meta* positions with respect to the substituent already present in the aromatic ring, may also be given .

Each of the three carbonium ions is a resonance hybrid of three contributing structures. The carbonium ions formed by electrophilic attack at *ortho* and *para* positions with respect to nitro group have one highly unstable contributing structure each. Relatively carbonium ion formed by *meta* attack is more stable. So *meta* attack is preferred when NO_2 or any other electron-withdrawing substituent is already present in the ring.

Case II—Ortho and para-directing groups

(i) *Ortho-and para -directing influence of OH in phenol,* for example, may be explained on the basis of three effects as below:

–I effect: The key atom oxygen of OH group in phenol has a strong -I effect and so tends to decrease the electron density mainly at *ortho* positions of the ring.

+E effect: In phenol, a lone pair of electrons in the p orbital of oxygen is in conjugation with the π electrons of the aromatic ring. When attacking agent is required ,+E effect is introduced into operation. This increases the electron density of the ring at *ortho* and *para* positions (Fig. 27).

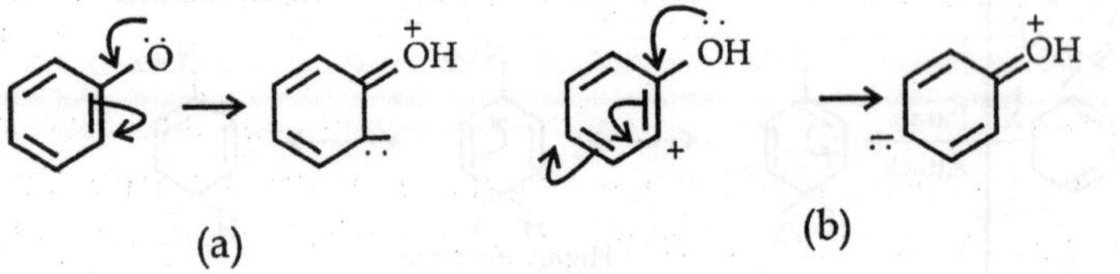

Fig. 27. +E effect of OH group at (a) ortho and (b) para position in phenol.

+M effect: The lone pair of p electrons on oxygen of the OH group in phenol, being in conjugation with the π electrons of the ring, is delocalised. So, eight π electrons are delocalised around the oxygen and the ring atoms. This gives rise to

five resonating structures of phenol (Fig. 24) three of which have negative - charged centres at *ortho* and *para* positions, indicating the increased electron density of the ring. This s +M effect of OH group.

Since +M and +E effects of OH group combined together are greater than the opposite-I effect, the ring of phenol is open for electrophilic substitution at o-and p-position mainly.

Note- The *ortho* and *para* directing influence of other groups, such as, $\ddot{O}R$, $-\ddot{N}H_2$, $-\ddot{N}H-\overset{O}{\overset{\|}{C}}-CH_3$, $-\ddot{O}-\overset{O}{\overset{\|}{C}}-CH_3$, etc, and the activation of the ring in their presence for electrophilic substitutions,can also be similarly explained.

(i)*O rtho-and para-directing influence of Cl in chloro-benzene* may also be explained on the basis of the three effects as below:

– I effect: Here it is very strong due to the strong electronegative character of Cl. Electrons are displaced towards chlorine, and the ring is *considerably deactivated.*

+M effect: In chlorobenzene ,the lone pair of p electrons on chlorine is in conjugation with the π electrons of the ring and, is thus delocalised around the ring atoms. Resulting in the displacing of electron from *Cl* to the ring . This is +M effect of chlorine, which gives rise to the five resonating structures of chlorobenzene. Three of these structures have negative-charged centres at *ortho* and *para* posintions:

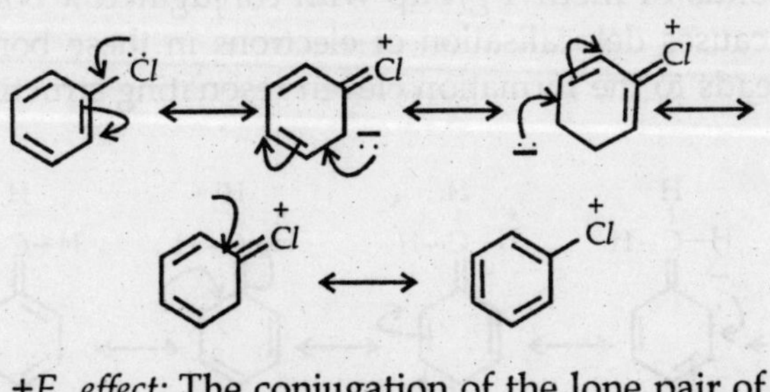

+E effect: The conjugation of the lone pair of p electrons on chlorine with the aromatic ring brings +E effect into operation, increasing the electron density at ortho and para positions like that in phenol.

Although, the +E effect supported by +M effect does not completely make up the loss of electron density caused by strong −I effect of Cl in the ring of chlorobenzene, yet on account of this effect, the electron density at ortho and para positions becomes higher than that at *meta* position.

Thus chlorine though deactivates the ring for electrophilic substitution, yet it remains ortho and para directing for such substitutions.Other halogens also behave like chlorine.

(iii) Ortho-and para-directing influence of methyl group in toluene may be explained on the basis of inductive, electromeric and hyperconjugative effects as below:

+I effect: Electron repelling nature of methyl group results +I-effect. This increases the electron density of the ring mainly at ortho positions.

+E effect: The +I effect of methyl group gives a positive direction to the electromeric displacement of π electrons of the ring creating negative charged centres at ortho-and para-positions (Fig. 28) at the requirement of the attacking reagent.

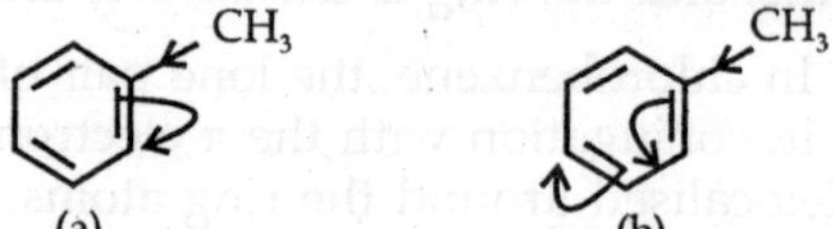

Fig. 28. +E effect of methyl group at (a) ortho and (b) para position in toluene.

Hyperconjugative effect – Hyperconjugation of three σ (C−H) bonds of methyl group with conjugated π bonds of the ring causes delocalisation of electrons in these bonds. This effect leads to the formation eleven resonating structures of toluene.

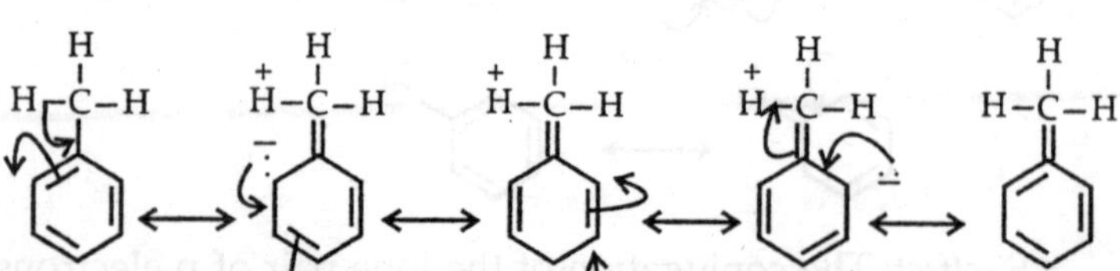

← → six more resonating structures involving the other two C−H bonds.

+I, +E and the hyperconjugative effect in positive

direction (+M) combined together are thus responsible for electrophilic substitution at ortho- and para positions in toluene and account for the activation of the ring.

Ortho and para directive influence of groups like $-OH$, NH_2, $-OR$, $-NHR$, $-X$ and $-CH_3$ may also be explained on the basis of relative stabilisations of intermediate carbonium ions formed by electrophilic attack of X^+ at various positions.

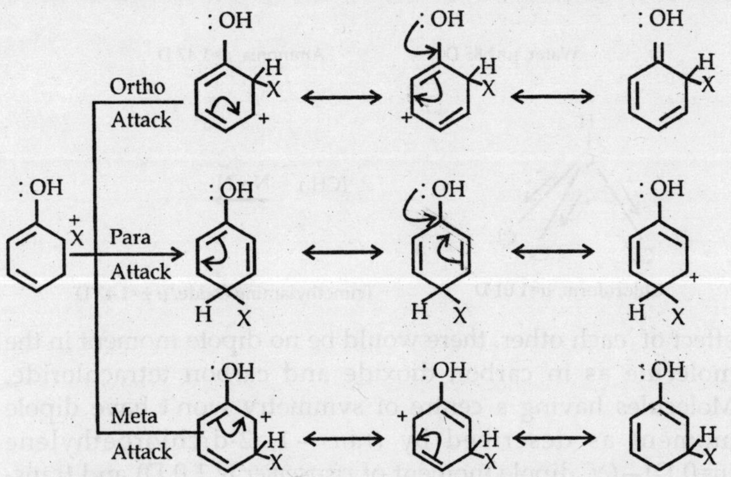

The carbonium ions formed by the *ortho* and *para* attacks have four resonating structures each, while that formed by *meta* attack has only three. Hence, the former are more stabilised Hence, in presence of activating groups as substituents in the ring, *ortho* and *para* electrophilic attacks are favourable.

DIPOLE MOMENT

Dipole is a pair of opposite charges of equal magnitude at a specific distance from each other. A dipole exists due to the concentration of positive charge at one and negative charge at the other end. The degree of polarity of a bond or molecule is measured by its dipole moment μ Dipole moments are reported in terms of Debye unit (D). 1D = 10^{-18} electrostatic units.

Molecules having polar bonds or formal charges possess dipoles. Water ,ammonia and chloroform having the smaller dipole moments illustrate the former, while trimethylamine oxide with a high dipole moment illustrate the latter type If the bonds in a molecule are so oriented as to cancel the polar

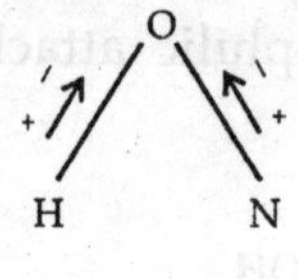

Water, μ=1.85 D

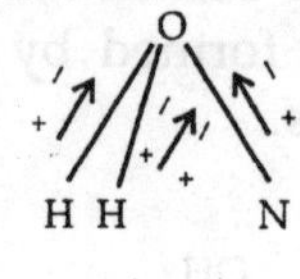

Ammonia, μ=1.47 D

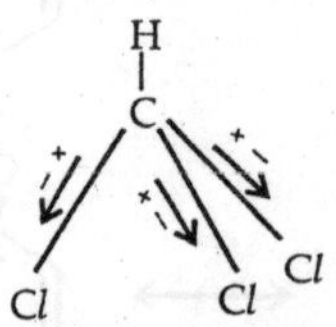

Chloroform, μ=1 01 D

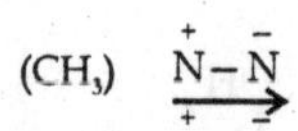

Trimethylamine Oxide, μ ±=1.47 D

effect of each other, there would be no dipole moment in the molecule as in carbon dioxide and carbon tetrachloride, Molecules having a centre of symmetry won't have dipole moment as described by *trans*−1, 2-dichloroethylene (μ=0 D)−(*cf.* dipole moment of *cis-isomer* is 1.0 D) and trans-2-butene (μ=0 D)−(*cf.* dipole momet of *cis*-isomer is 0.33 D)

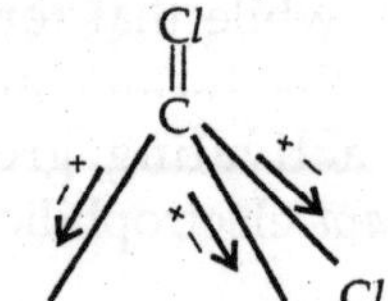

Tarbontetrachloride
μ=1 01 D

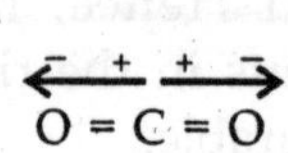

Carbondioxide,
μ=0 D

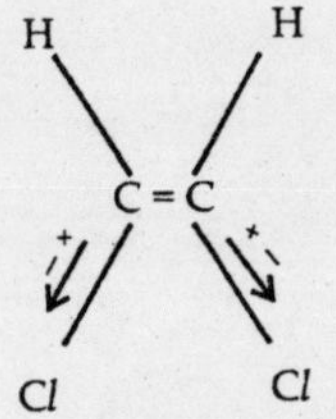

Cis – 1, 2 – dichloroethylene
μ=1 D

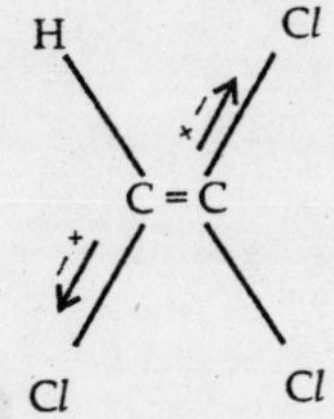

Trans – 1, 2 – dichloroethylene
μ=0 D

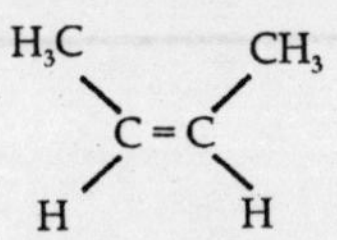

Cis – 2 butene
μ=0.33 D

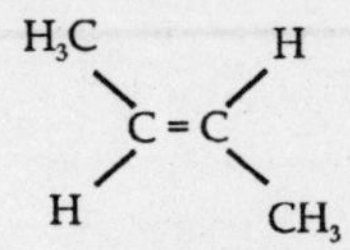

Trans – 2 – butene
μ=0 D

Conjugation which leads to resonance also affects the dipole moment, as can be seen from the dipole moments of nitromethane and nitrobenzene. In the latter, NO_2 group is conjugated.

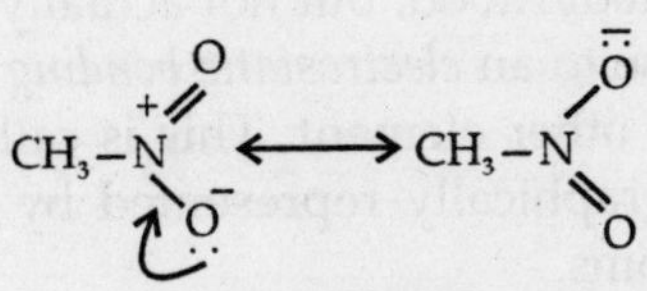

Nitromethane,
μ=3.5 D

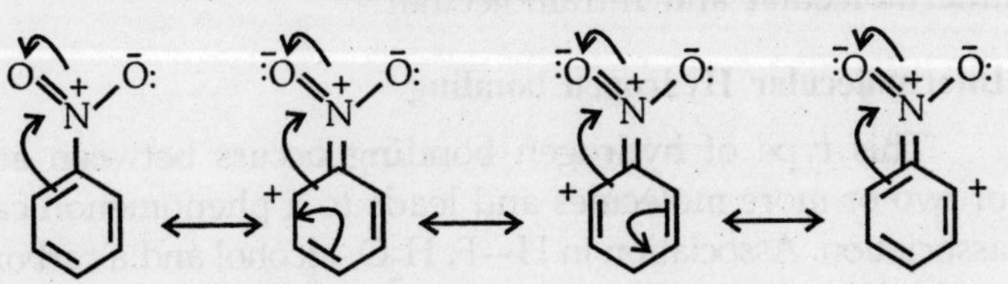

Nitrobenzene, μ=3.95 D

Chapter 5

HYDROGEN BONDING

Z–H bond is polarised if Z is highly electronegative atom of small atomic radius, such as *fluorine, oxygen* and *nitrogen**. Due to polarisation, hydrogen atom has a large fraction of unit positive charge (δ^+) and Z has an equal fraction of negative charge (δ^-) in the molecule Z–H Since hydrogen atom has the smallest volume, the δ^+charge of H exerts a great electrostatic force of attraction of the highly electronegative atom of the above-mentioned elements present in the neighbourhood, but not actually bonded with this H. This gives rise to an *electrostatic bonding* between Hδ^+ and the atom of the other element. This is called Hydrogen bonding, which is graphically represented by a dotted line between the two atoms.

There are two types of Hydrogen bonding Intermolecular and Intramolecular.

Intermolecular Hydrogen bonding

This type of hydrogen bonding occurs between atoms of two or more molecules and leads to a phenomenon called association. Association in H–F, H_2O, alcohol and a carboxylic acid is described in Fig. 29.

* With atoms of greater atomic radii, the electrostatic forces of hydrogen-bonding are weaker. Chlorine, for example, has the same electronegativity as nitrogen, but, because of its greater atomic radius it forms very weak hydrogen bonds.

Intramolecular hydrogen bonding

This type of hydrogen bonding occurs between the atoms of the same molecule and results to a phenomenon called chelation.

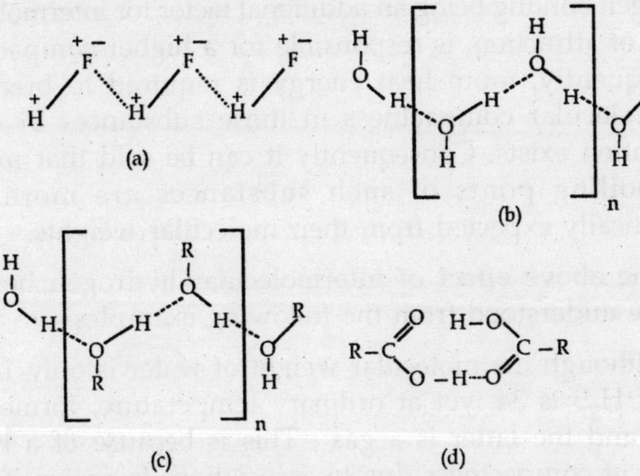

Fig. 29 Intermolecular hydrogen bonding and association in (a) hydrofluoric acid; (b) water; (c) alcohol; (d) carboxylic acid. (n is an integer)

Chelation in *o*-nitrophenol, salicylaldehyde and salicylic acid is illustrated in Fig.30.

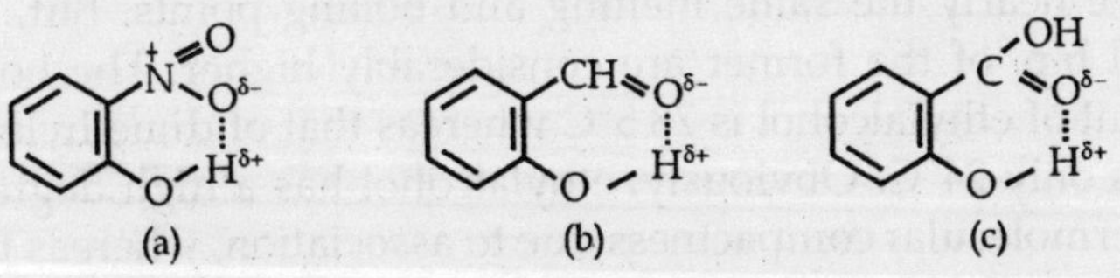

Fig. 30 Intramolecular hydrogen bonding and chelation in (a) *o*-nitrophenol; (b) salicylaldehyde; (c) salicylic acid.

HYDROGEN BONDING AND ITS EFFECTS

Hydrogen bonding affects physical properties such as melting point boiling point and solubility.

(a) Effect on melting and boiling points—State of a substance *i.e.* it is solid, liquid or gaseous depend on its intermolecular compactness, that is due to intermolecular forces of attraction. Greater the intermolecular forces of attraction, more is the compactness. For a solid to melt or a liquid to vapourise, energy in the form of heat is required to weaken the intermolecular forces of attraction thereby reducing the compactness. Association of molecules due to hydrogen bondng being an additional factor for intermolecular forces of attraction, is responsible for a higher compactness. Consequently, more heat energy is required to break the intermolecular compactness in those substances in which association exists. Consequently it can be said that melting and boiling ponts of such substances are more than theoretically expected from their molecular weights.

The above effect of intermolecular hydrogen bonding may be understood from the following examples–

Although the molecular weight of water is only 18 and that of H_2S is 34, yet at ordinary temperature, former is a liquid and the latter is a gas . This is because of a higher degree of compactness due to association in water. That is why, the melting and boiling points of H_2O are considerably higher than those of H_2S in which there is no intermolecular hydrogen bonding.

The molecular weight of ethylalcohol and dimethylether is the same (46), and so both are theoretically expected to have nearly the same melting and boiling points. But, m.p. and b.p. of the former are considerably higher. The boiling point of ethylalcohol is 78.5°C whereas that of dimethylether, it is only-24°C. Obviously, ethylalcohol has a high degree of intermolecular compactness due to association, whereas there is no association in dimethylether.

The effect of intramolecular hydrogen bonding i.e. chelation is opposite to that of association. This phenomenon weakens the intermolecular forces of attraction, and reduces the intermolecular compactness of a substance . Hence, to break this compactness, less heat energy is required i.e. the melting and boiling points of such substances are lower than theoretically expected from their molecular weight. For example, the molecular weight of o⁻ m⁻ and p-nitrophenols is the same (139), but the melting pont of o⁻ nitrophenol is considerably lower than that of the other two. This is because of a high degree of chelation in *o*-nitrophenol.

(b) Effect of hydrogen bonding on solubility– Hydrogen bond formation between the molecules of different substances is called co-association. When co-association takes place between the molecules of a *solute* and *water*, it brings about an intimate dispersion of the solute molecules between the water molecules, but this is possible only when the solute molecules have a polar group to break the association of water (solvent) molecules. This ability of solute to co-associate with water molecules due to hydrogen bonding is responsible for its solubility in water. Alcohol is soluble in water due to the molecules of the two:

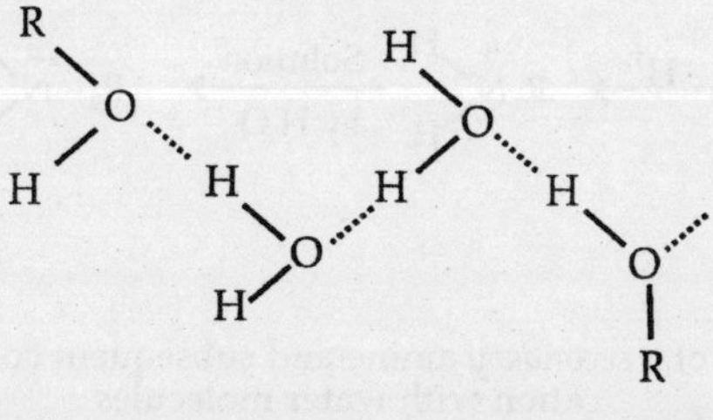

Solubility of hydroxy compounds in water depends upon the availability of OH groups in their molecules for co-association withe water molecules. The availability of OH groups decreases with increase in the size of alkyl group in alcohols (R–OH). So the higher alcohols are insoluble in water.

In case of polyhydroxy compunds (sugars), the large number of OH groups in the molecule are available for co-association with water molecules. Consequently, such compounds are highly soluble in water.

At room temp, primary and secondary amines, possessing not more than 5 and 4 carbon atoms respectively, are completely miscible with water. This is because their nitrogen atom easily accepts the proton from water (protonation) to form cation, and the latter co-associates water molecules due to hydrogen bonding.

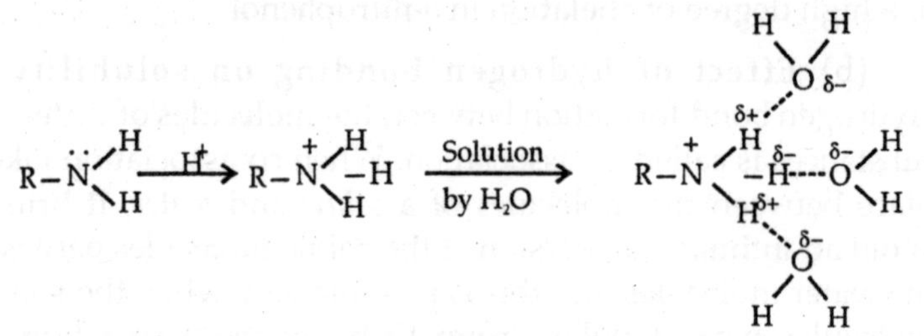

Protonation of a *primary* amine and subsequent co-association of the cation with water molecules.

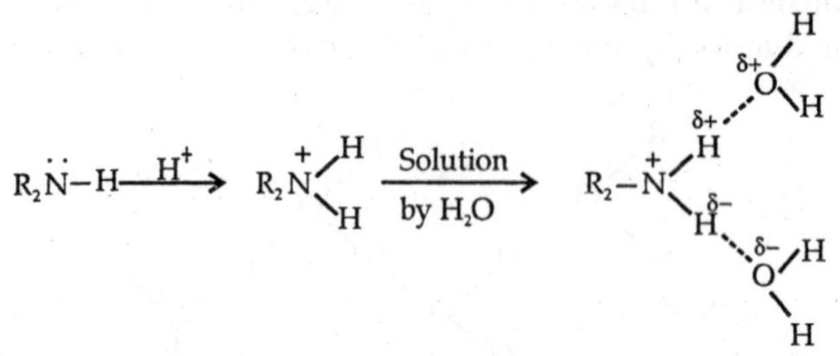

Protonation of a *secondary* amine and subsequent co-association of cation with water molecules .

In *tertiary* amines, the presence of three alkyl groups on nitrogen atom, creates *steric hindrance* for protonation. Hence, in these amines ,the formtion of cation is difficult. The extent of *steric hindrance,* however, depends upon the size of the alkyl groups. Since in the simplest *tertiary* amine—trimethylamine—the alkyl groups are small, there is some possibility of protonation, and therefore this amine is soluble to the exent of 41%. As the size of alkyl groups increases, the solubility of *tertiary* amines decreases. Other *tertiary* amines are, therefore, insoluble.

Intramolecular hydrogen bonding (chelation) in solute molecules is responsible for reduced polarity in these molecules. So the molecules of such solute fail to break the intermolecular association of water. Chances of co -association between the molecules of the solute and water are thus reduced and the solubility of substance in water decreases. For example, m-and p-nitrophenols are more soluble in water than o-because of chelation in the molecules of the latter.

General rules of solubility

Solubility of a substance is affected by intermolecular forces existing between (a) the molecules of the solute, (b) the molecules of the solvent, and (c) the molecules of the solvent and the solute "Like dissolves like" is a general principle of solubility in organic chemistry. It can be concluded that non-polar solutes are not soluble in polar solvents. but are soluble in non-polar sovents .Similarly polar solutes are insoluble in non-polar solvents, but are soluble in polar solvents.

1. Non-polar solute and polar solvents

Molecules of a polar solvent, such as water, normally remain in associated state. Hence, strong intermolecular forces of attraction exist between these molecules, The molecules of the nonpolar solute are, therefore, unable to break the intermolecular compactness of the solvent and are, thus insoluble. So, due to nonpolar nature of the hydrocarbon groups, the hydrocarbons and other organic compounds having larger hydrocarbon group such as ethers, higher alcohols, higher acid, higher amines and esters are insoluble in water.

2. Non-polar solute and non-polar solvents

Non-polar solutes are soluble in non- polar solvents, because the intermolecular forces of attraction between the molecules of each are very weak due to lack of association. In such cases, the solubility is rarely a free intermingling or free dispersion of solute and solvent molecules.This is how, non-polar solutes (hydrocarbons and others organic compounds),with large hydro-carbon group, dissolve in non-polar solvents (ether, benzene,toluene etc.) having predominant hydrocarbon portion in their molecules,

3. Polar solute and nonpolar solvents

Polar solutes such as salts and other compounds having large number of polar groups (sugars, ethylenediamine etc.)do not dissolve in nonpolar solvents. The nonpolar solvents have low dielectric constant (or low polarising power) due to which they are unable to ionise the polar solutes and thus, fail to dissolve them. In addition to this, the non-polar solvent molecules are unable to form hydrogen bond with the polar solute molecules , and thus the dissolution of solute molecules by co- association also fails to occur.

4. Polar solute and polar solvents

Polar solute dissolves in polar solvent, because–

(i) Intermolecular forces of attraction between the polar ends of solvent and solute molecules,called ion-dipole bonds, are stronger than the forces of association of solvent molecules. The association of solvent molecules is thus broken.

(ii) Each *ion-dipole bond* is relatively weak, but in aggregate these are strong enough to overcome the interionic electrostatic attraction in the solute molecule. Hence two ions of the solute molecules are separated and shielded by the solvent molecules resulting into dissolution.

Dissolution of sodium chloride in water ,as illustrated below, is an example of such a case–

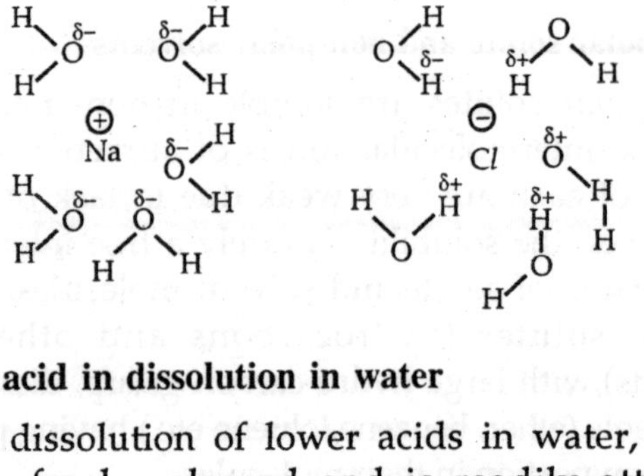

Carboxylic acid in dissolution in water

In the dissolution of lower acids in water, proton of the acid is freely released and is readily attracted by negative oxygen of H_2O to form H_3O^+ ion, whereas carboxylate ion of the acid dissolves by forming hydrogen bond with positive end of H_2O In case of higher acids, which have large alkyl group in their molecules, the proton remains a captiveof the alkyl group. This the release of proton is hindered sterically, fatty acids cannot coassociate with water and are insoluble. Sodium salts of higher acids, however, dissolve in water. Thus is due to the larger size of Na^+ as compared to H^+, the former being free enough to be released (by ionisation of the salt) for co-association with water molecules.

Chapter 6

STEREOISOMERISM

The isomerism that arises out of different orientations of atoms or groups (the respective attachment of the atoms remaining the same)in space is referred to as stereoisomerism. So, stereoisomers *i.e.* spatial isomers are isomeric compound which have same molecular formula but differ in the arrangement of atoms or groups in 3–D space. Stereoisomerism is of two types- (i) **Optical isomerism** and (ii) **geometrical isomerism.**

OPTICAL ISOMERISM

Most of the properties of light are explained if it is deemed to be a wave. Oscillations in the wave of a ray of *ordinary* light take place in infinite number of planes perpendicular to the direction of propagation (Fig 31 a) The ordinary light

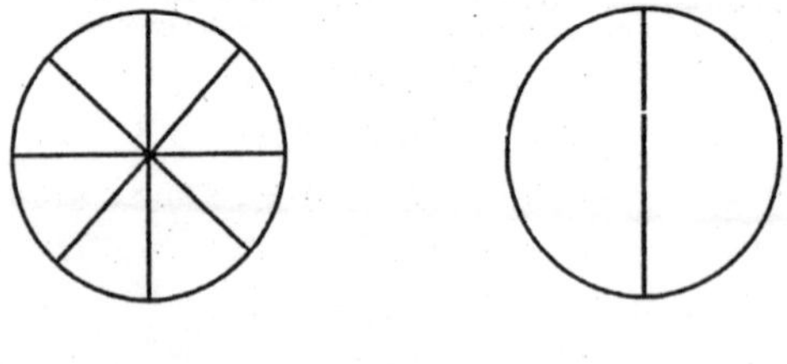

(a) (b)

Fig. 31 – (a) Oscillations in an ordinary light (b) oscillations in a planepolarised light.

which when passed through a Nicol prism emerges out as plane polarised light in whch oscillations occur only in one plane . Substances , which are capable of rotating the plane of polarised light are said to be **optically active**. The property of rotating the plane polarised light is called **optical activity**. Polarimeter is the device which is used to determine and measure the optical activity of a substance in terms of angle of rotation of the plane in degrees. If the rotation is towards right (clock-wise), the substance is **dextrorotatory** and if it is towards left (*anticlock-wise*), the substance is levorotatory. To denot dextro-and levorotation, symbols used are (+) and (–) respectively.

The extent of optical rotation (angle) is directly proportional to the number of molecules of the substance light has to encounter with i.e. on concentration and the path so, quantitatively, the rotation is expressed in terms of specific rotation by the Specific rotation (α) is defined as the degrees of rotation shown by the substance of 1gm/ml concentration taken in l decimeter long tube. Thus:

$$[\alpha]_{\text{Wavelength}}^{\text{Temp.}} = \frac{\text{Observed rotation (degrees)}}{\text{Length of the tube in dm} \times \text{concn. (g/ml)}}$$

The only factor responsible for the rotation of the plane of polarised light is **molecular asymmetry**. An object which has no plane of symmetry (a plane which bisects the object into two super-impossible mirror image halves is called the plane of symmetry) is asymmetric. The reflection of an asymmetric object in a mirror is not superimposable on the object. Asymmetric object and its mirror image are called **enantiomers**. A pair of gloves illustrates this phenomenon. The two gloves are mirror images of each other but are not superimposable. The pair of human body and its mirror image is another example in which the left part is reflected in the mirror as right part, and vice versa. In the same way to asymmetric molecules that are superimposable mirror images of each other are two different isomeric entities (enantiomers) differing only in the arrangement of atoms or groups. These isomers are identical in all their physical and chemical properties except in the direction of rotation of the plane of polarised light. If one of the pair rotates the plane of polarised light towards right, the other rotates towards left to the same extent. By virtue of possessing optical activity, the enantiomers are also called **optical isomers.**

As we have seen in Chapter 2, the sp^3 hybridised carbon atom is tetrahedral. The four substituents linked to a carbon atom are directed towards the vertices of a regular tetrahedron (Fig.9). If all the four substituents on a carbon atom are different—say C (PQRS), then its mirror-image does not superimpose it (Fig.32). Such a carbon atom is called an **asymmetric carbon atom**. Conversely if any of the two subtituents are identical, the mirror-image superimposes as can be seen in fig 33 and doesn't give rise to asymmetry. Superimposition and non-superimposition can be easily understood by the help of models.

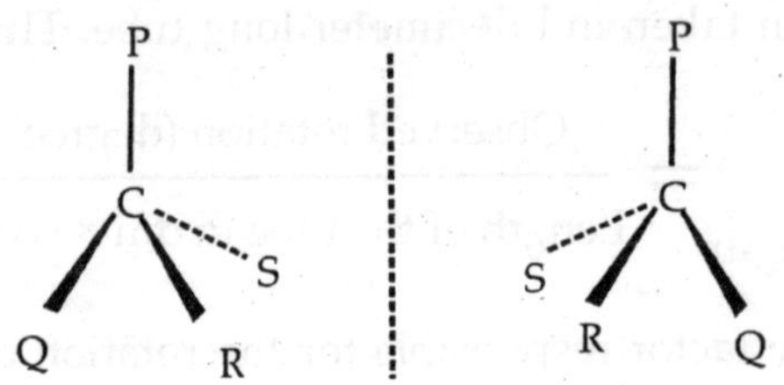

Fig 32—Non-superimposable mirror-images.

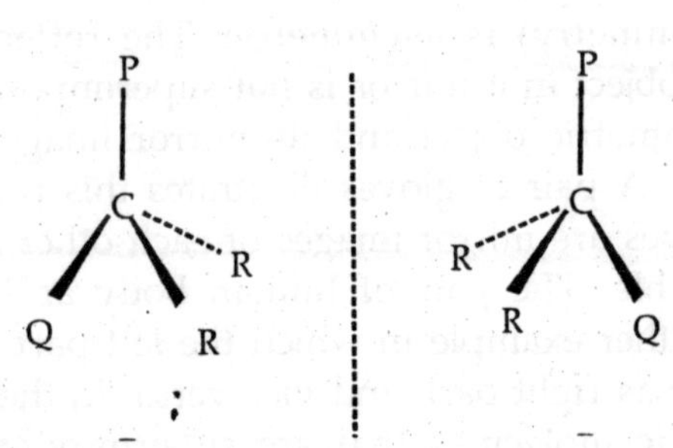

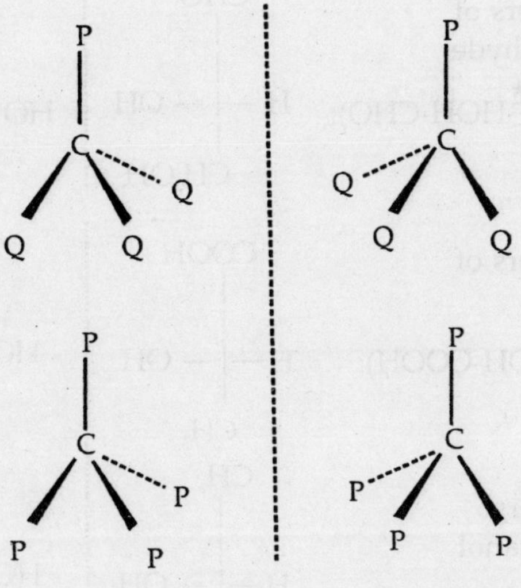

Fig. 33—Superimposable mirror-images

An asymmetric carbon atom brings. asymmetry in the molecule, a factor responsible for optical activity. Thus for a compound to be optically active, it is essential that it must possess an asymmetric carbon centre. The reverse on the other hand is, however, not true i.e. the presence of an asymmetric carbon centre in a molecule does not compulsorily make it optically active(see the case of optically inactive meso-tartaric acid, which possesses two asymmetric carbon atoms, but there lies a plane of symmetry).

Molecules which have one asymmetric carbon atom ($\overset{*}{C}$) exist in two optically active forms—one being the non-superimposable mirror image of the other (enantiomeric pair). Enantiomeric pairs of some of the optically active compounds are given below. Horizontal lines in these Fischer projection formulas represent the bonds coming out of the plane of parper, while the vertical lines represent bonds going behind the paper:

Enantiomers of glyceraldehyde ($HOCH_2$-$\overset{*}{C}$HOH-CHO):

Enantiomers of lactic acid (CH_3-$\overset{*}{C}$HOH-COOH):

Enantiomerso 1-phnylethanol (ph. $\overset{*}{C}$HOH-CH_3):

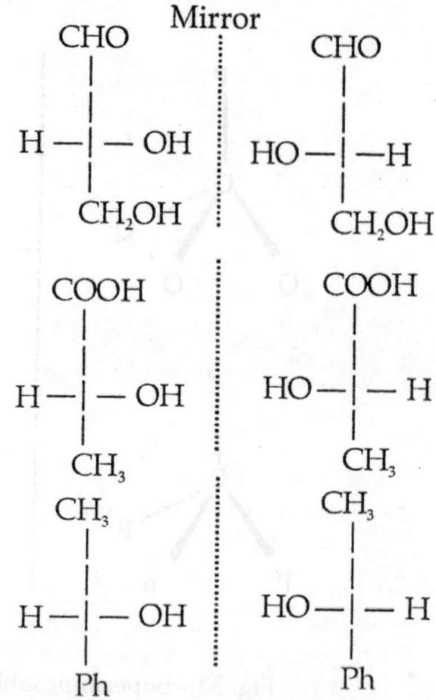

The eqimolar mixture of an enantiomeric pair becomes optically inactive as a result of the cancellation of the equal and opposite rotations of the isomers. Such a mixture is termed as racemic mixture of modification. The symbol ± is used to represent the racemic nature of a sample. Most of the physical properties of enantiomers being identical, it is impossible to separate them by usual methods like fractional crystallization, distillation or chromatography. Enantiomers are always formed in equal amounts in laboratory syntheses and the product is thus a racemic mixture. The separation of the optical isomers from a recemic mixture. The separation of the optical isomers from a recemic mixture, a process called *resolution,* can be accomplished by converting the isomers *temporarily* into derivatives that no longer remain mirror images. These derivatives have different solubilities and which can be separated by crystallization.

Absolute configuration: The spatial arrangement of four different atoms or groups around the asymmetric carbon atom which specifies an enantiomer is called its **configuration**. The *absolute configuration* characterises the order of the substituents in such a way as to distinguish the two enantiomers. D–and C–prefixes are used to recognise the absolute configuration of an enantiomer. The dextrorotatory (+) enantiomer of glyceraldehyde has been arbitrarily assigned the configuration shown in Fig. 34 (a) and is designated as 'D', while its mirror image, the levo-rotatory (–) enantiomer as 'L'. Enantiomers of other

(A) (B)

Fig 34–(a) Configuration of D(+)–glceraldehyde and (b) configuration of L(–) glyceraldehyde.

substances are assigned D or L configurations according to configurational similarity with D-or L- glyceraldehyde.

The configuration of all α-hydroxy acids and α-amino acids that contain OH/NH_2 on the right hand side in the Fisher projection formula are designated by D. If the OH/ NH_2 is on the left, L is used. In carbohydrates, for determining the desingnation, asymmetric centre at the bottom is taken into consideration.

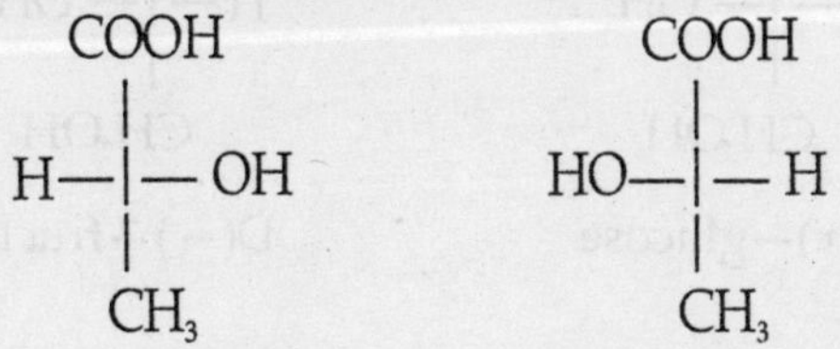

D(–)–lactic acid L(+)–lactic acid

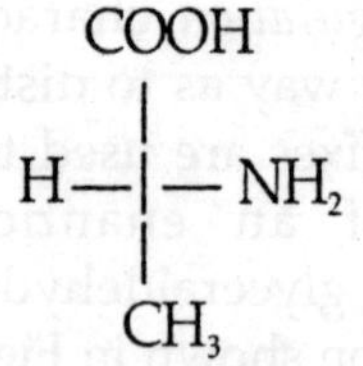

D(–)–alanine

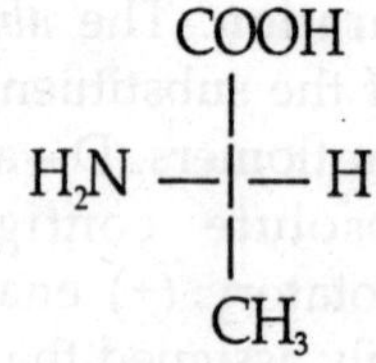

L(+)–alanine

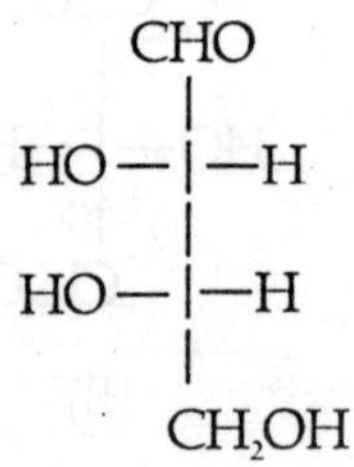

L(+)–erythrose

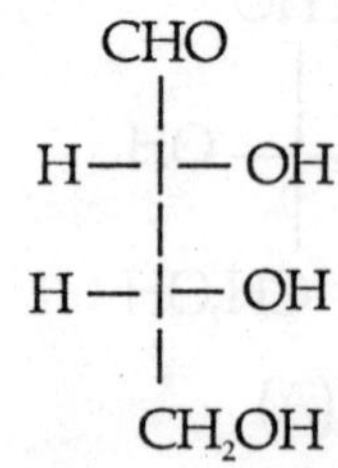

D(–)–erythrose

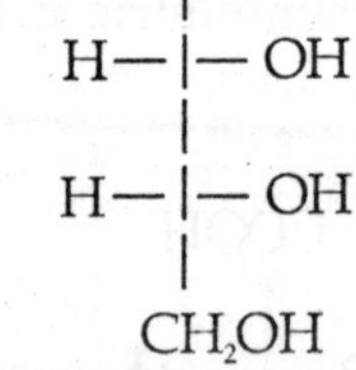

D(+)–glucose

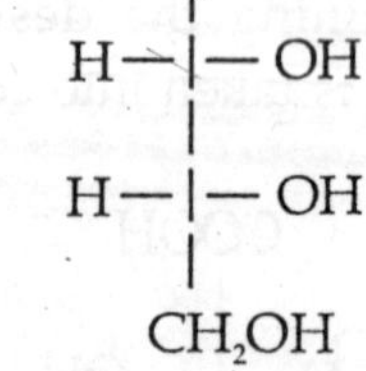

D(–)–fructose

It must be borne in mind that the rotations (+) or (–) have nothing to do with these configurations. An enantiomer may be D(+) or D(–) and likewise L(+) or L (–).

So far we have taken the examples of compounds having only one asymmetric carbon atom more than one asymmetric carbon atom at the centres, the number of stereoisomers are given by 2^n, where n is the number of asymmetric carbon atoms.

2, 3-dichloropentane, $\underset{5}{CH_3}\underset{4}{CH_2}-\underset{\underset{Cl}{|}}{\overset{*}{\underset{3}{C}}H}-\underset{\underset{Cl}{|}}{\overset{*}{\underset{2}{C}}H}-\underset{1}{CH_3}$ has two asymmetric carbon atoms and exists in four stereoisomeric forms:

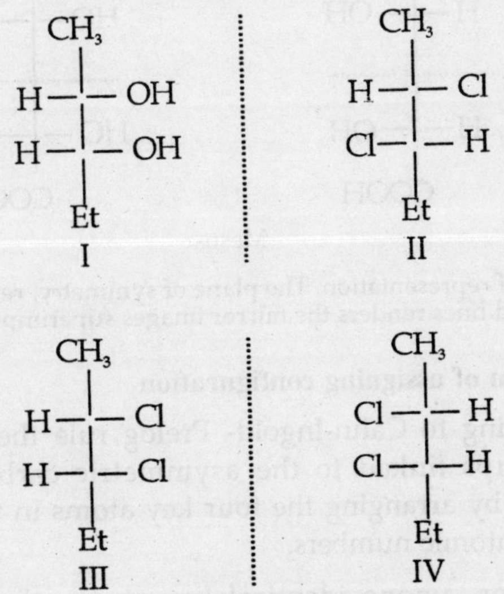

(I and II) and (III and IV) are enantiomeric pairs. (I and III), (I and IV) and (II and III) are diastereomrs.†

Tartaric acid $(COOH-\overset{*}{C}OOH-\overset{*}{C}OOH-COOH)$ has three stereoisomers, out of which two are optically active and one is optically inactive.

---

† Diastereomers are the stereoisomers that are not the mirror images of each other.

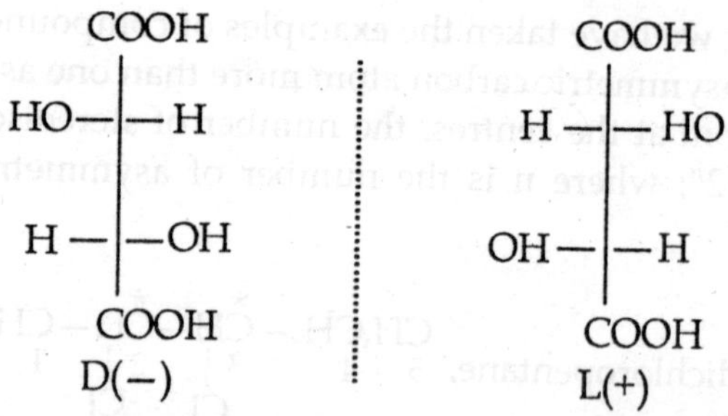

Two enantiomers, the non-superimposable mirror mages

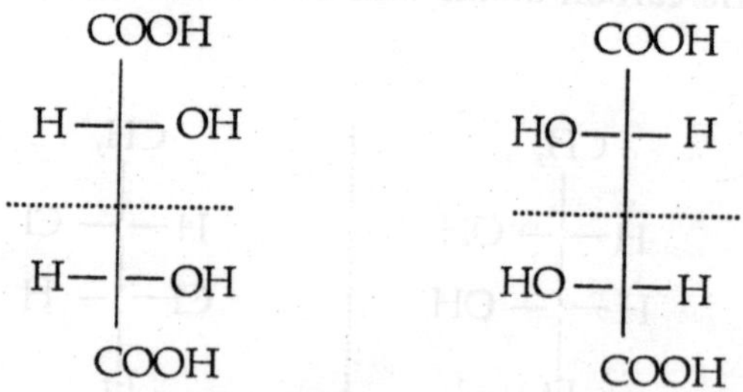

Meso-

Two ways of representation. The plane of symmetry, represented by dotted lines renders the mirror images superimposable

R, S—system of assigning configuration

According to Cahn-Ingold- Prelog rule the priority of atoms/groups linked to the asymmetric carbon atom is ascertained by arranging the four key atoms in the order of decreasing atomic numbers.

However, among identical key atoms, the priority is decided by taking into account the atomic numbers of the subsequent atoms linked to them. For this purpose, double-bonded atoms such as C = C are deemed as 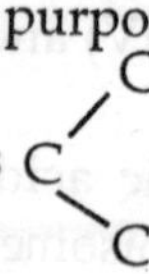 and C = O as 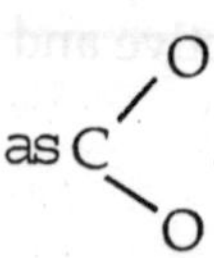 . Similarly C≡C taken as 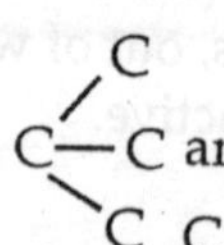 and C≡N as 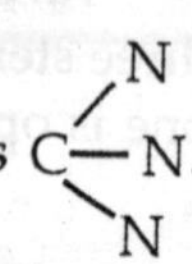.

The phenyl group is deemed as 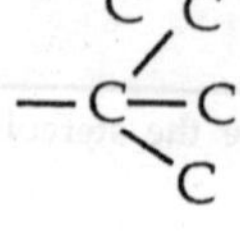 Examples in the following table illustrate the fixation of priority of atoms and groups bonded to the asymmetric carbon atom in optically active compounds:

Structure of the optically active compound	Order of priority of atoms and groups
$CH_3CH_2-\overset{*}{C}H(Cl)-CH_3$	$Cl > CH_2CH_3 > CH_3 > H$
$CH_3-CH(CH_3)-\overset{*}{C}H(Cl)-CH_2CH_3$	$Cl > -CH(CH_3)CH_3 > CH_2CH_3 > H$
$HOCH_2-\overset{*}{C}H(OH)-CHO$	$OH > CHO > CH_2OH > H$
$CH_3-\overset{*}{C}H(NH_2)-COOH$	$NH_2 > COOH > CH_3 > H$
$C_6H_5-\overset{*}{C}H(NH_2)-CH_3$	$NH_2 > C_6H_5 > CH_3 > H$

Now, the tetrahedral asymmetric carbon atom is observed so that the atom or group of lowest priority is found out. The remaining three groups are then looked. If the eye has to move in a clock-wise direction in going from the group of highest priority to that of second priority and then to the third, the configuration is specified as R–(*rectus* = right); if in anticlockwise, it is specified as S–(*sinister* =

left) For example, *sec.* butylchloride $(CH_3CH_2-\underset{\displaystyle Cl}{\underset{|}{\overset{*}{C}H}}-CH_3)$ has H as a substituent of lowest priority. The molecule is viewed so as to keep H pointing away. Now arrange the remaining two groups in two way in the two enantiomers. In going from $Cl \rightarrow CH_2CH_3 \rightarrow CH_3$, Fig. 35 (a), the sequence is clockwise and hence the configuration is R, while in Fig. 35 (b), this sequence is anticlockwise with S−configuration.

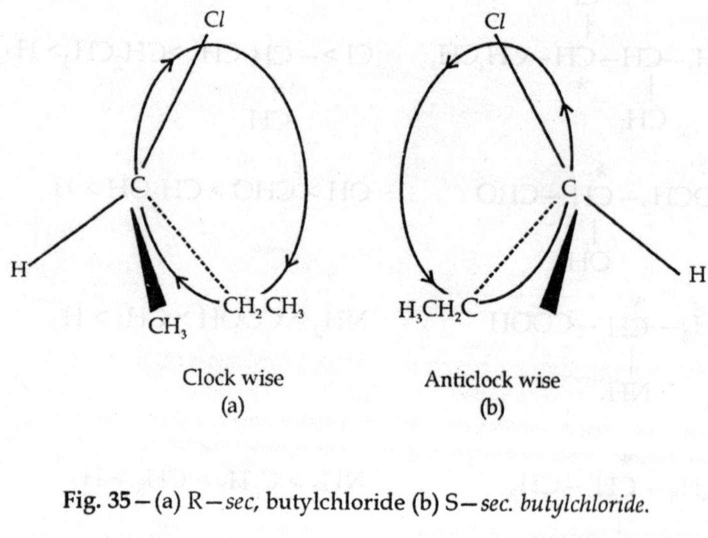

Fig. 35—(a) R−*sec*, butylchloride (b) S−*sec. butylchloride.*

In molecules, where there are two of more asymmetric carbon atoms, each of them is taken in to account R-or-S-configuration is assigned. The stereoisomers I, II, III and IV of 2, 3-dichloropentane are thus assigned (2R, 2R), (2S, 3S,) (2S, 3R and (2R, 3S) configurations respectively.

---

* 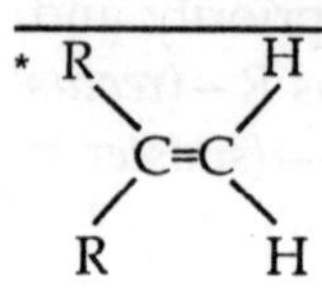 would not give rise to isomers by changing the spatial arrangement.

GEOMETRICAL ISOMERISM

Geometrical isomerism of ethylenic compounds

The carbon atoms linked by a double bond are sp_2-hybridised. σ and π MOs together constitute a double bond. The presence of perpendicular π MO restrict a free rotation of carbon nuclei about the double bond (energy barrier=63 kcal/mole) gives rise to the phenomenon of **Geometrical isomerism** or Diastereomerism. If spatial geometry of substituents attached to the double-bonded atoms is altered two stereoisomers are possible but remember none of the double-bonded atoms carries the two like-substituents.* These stereoisomers are called *diasteremers* (because they are not the mirror images of each other) or simply *geometrical isomers.* The pair of these isomers are referred to as *cis* and *trans.*

The *cis* isomer is the one in which the like-substituents

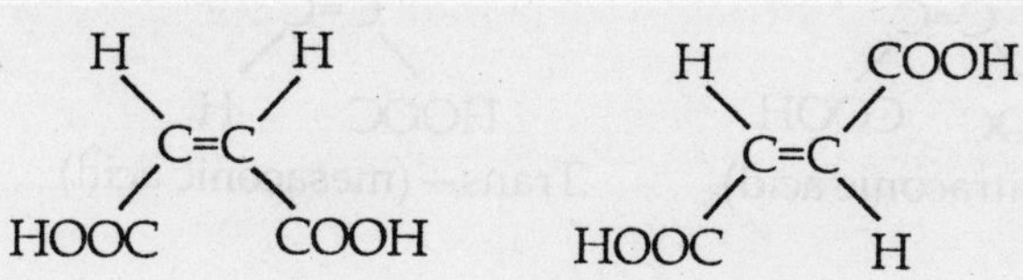

Cis (maleic acid) trans (fumaric acid)

are on the same side (*e.g.* maleic acid) and the isomer is *trans* if they are on the opposite sides (*e.g.* fumaric acid). If four different substituents are linked to double-bonded atoms. (*e.g.* in 3-methyl-3-heptene), *cis* isomer is the one in which bulkiest groups are on the same side:

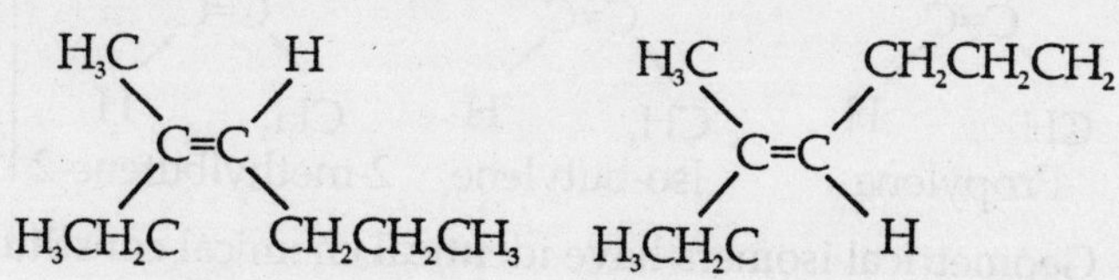

Cis—3-methyl-3-heptene Trans—3-methyl-3-heptene

A few examples are given below to illustrate:

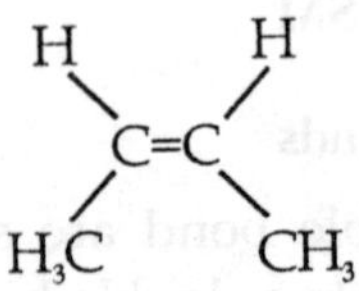
Cis – 2-Butene

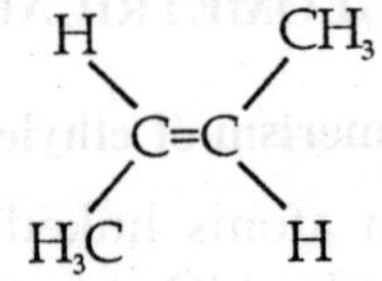
Trans – 2-Butene

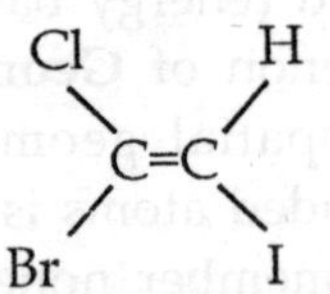
Cis – 1-bromo-I-chloro 2-iodoethylene

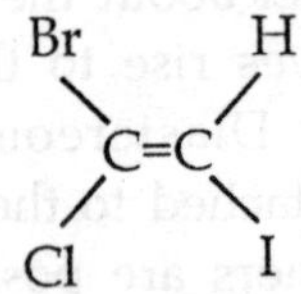
Trans – 1-bromo-1-chloro-2-iodoethylene

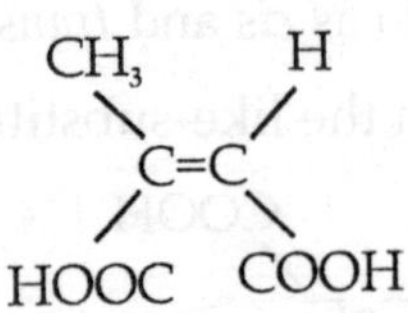
Cis-(citraconic acid)

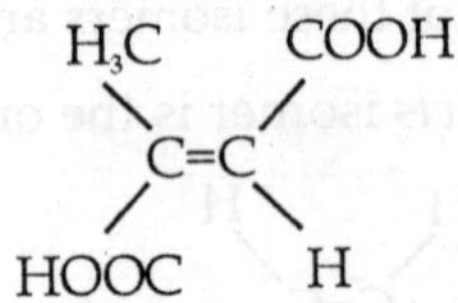
Trans – (mesaconic acid)

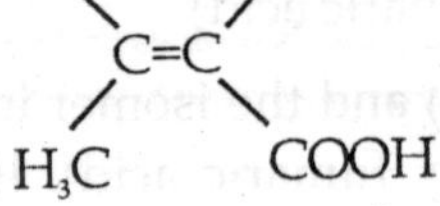
Cis-(crotonic acid)

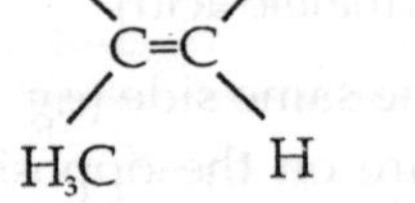
Trans – (iso-crotonic acid)

Geometrical isomeric pairs

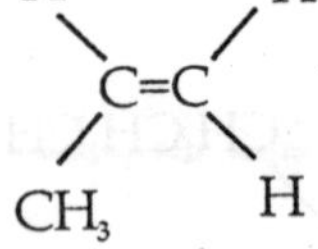
Propylene

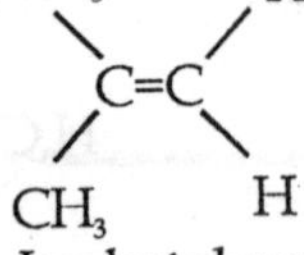
Iso-butylene

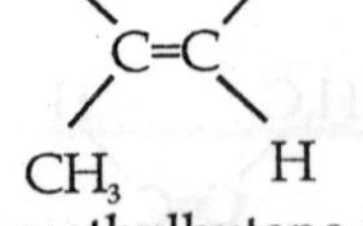
2-methylbutene-2

No geometrical isomerism

Geometrical isomers have identical chemical constitution and differ in the arrangement of atoms in space *i.e.* in configuration. It is this differecnce in configuration which is responsible for different chemical and physical properties of geometrical isomers. These isomers exist as *different* identities and can be isolated. Trans isomers have greater symmetry and hence more compactness of molecules. For example, maleic acid (*cis* isomer), m.p. 130°, fumaric acid (*trans* isomer), m.p. 287°; cinnamic acid, *cis*, m.p. 68°; *trans*, m.p. 133°; but-2-enoic acid, *cis* (isocrotonic acid), m.p. 15°; *trans* (crotonic acid), m.p. 72°; heptadec-8-ene-1-carboxylic acid, *cis*, (oleic acid), m.p. 16°; *trans* (elaidic acid), m.p. 51°; but-2 ene, *cis*, m.p. −139°; *trcns*, m.p. −106°. These isomers interconverse when heat and. UV radiation is used.

Another nomenclature for geometrical isomers recommended by IUPAC is *Z*, *E*-system. taken from German words Entgegen and Zusammen which means 'opposite' and 'together' respectively. In this system, configuration Z is assigned to that geometrical isomer which has substituents of higher priority on the same side of the plane of C=C and E to that which has substituents of higher priority across. The priority of the substituents is determined on the basis of atomic numbers of key atoms in decreasing order as in R, S-system of naming optical isomers (page 54).

A few examples are given below for illustration:

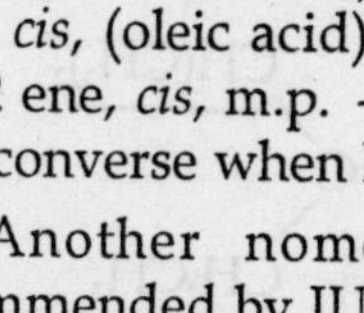

Z-1, 2-dichlo-1-propene (substituents of higher priority in each pair are on the same side)

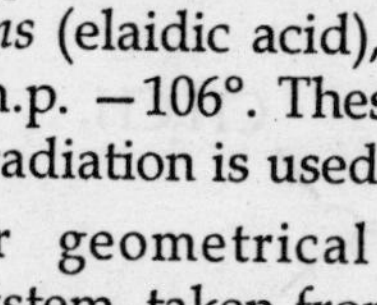

E-1-bromo-1chloro-2-nitroethane (substituents of higher priority in each pair are across)

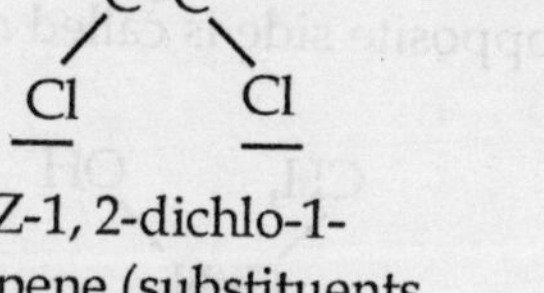

Z-but-2-ene-1, 2-dioic acid

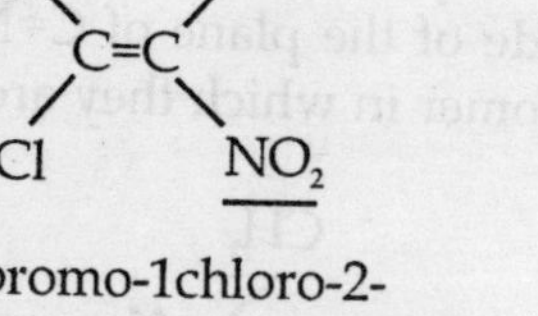

E-2phenylethene-1-carboxylic acid

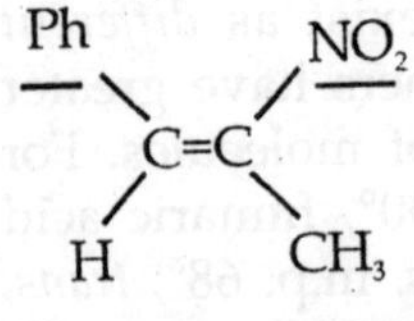

Z-1-phenyl-2-nitropropene

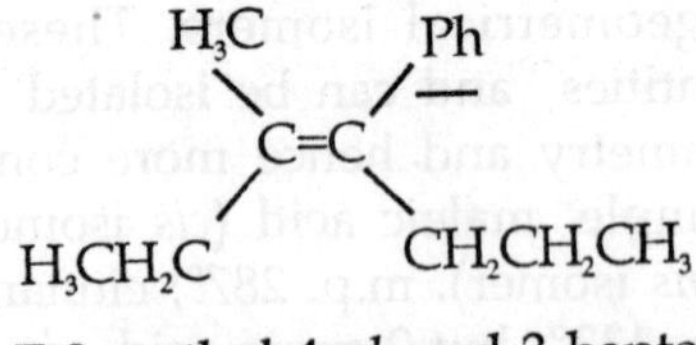

E-3-methyl-4-phenyl-3-heptene

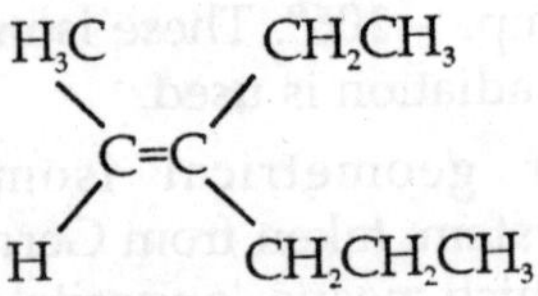

E-3-ethylhexene-3

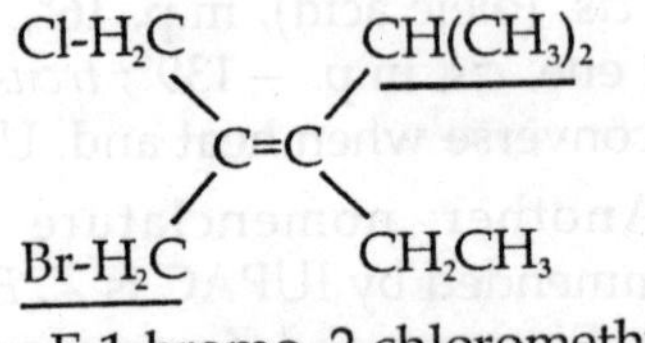

E-1-bromo, 2-chloromethyl, 3-ethylhexene-2

Geometrical isomerism of oximes and azo compounds

In analogy to *cis-trans* isomerism of ethylenic compounds, oximes and azo compounds which have C=N and N=N linkage respectively exist as two geometrical isomers called *syn* and *anti*.

The geometrical isomer of an aldoxime (oxime of an aldehyde *i.e.* RCH=N–OH) having H and OH on the same side of the plane of C=N is referred to as *syn* and the other isomer in which they are on the opposite side is called *anti*.

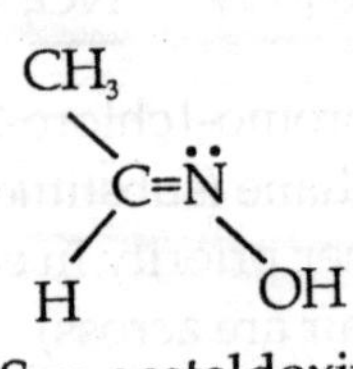

Syn-acetaldoxime

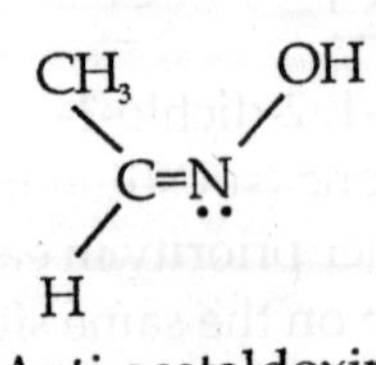

Anti-acetaldoxime

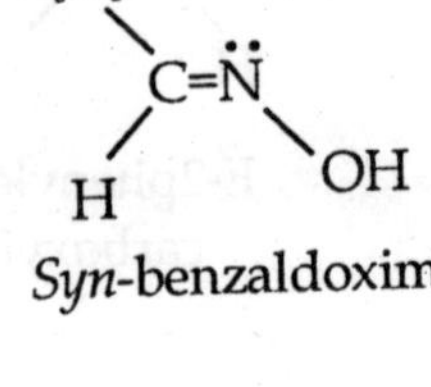

Syn-benzaldoxime

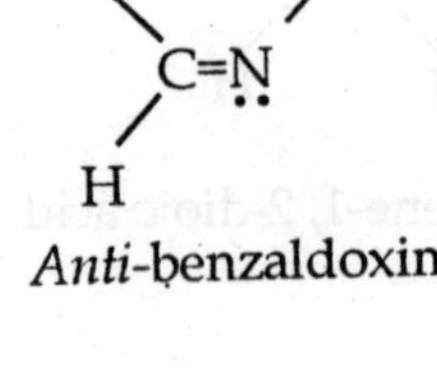

Anti-benzaldoxime

The oxime of a symmetrical ketone (*e.g.* 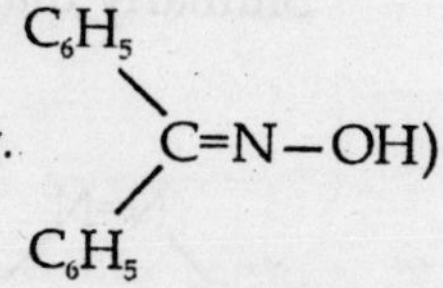)

has no geometrical isomers. At carbon of C=N, if two substituents are different the oximes exists in two forms but it is always not feasible to use the prefixes *syn* and anti to designate them. For such isomers, Z, E- system of nomenclature based on the priority of the substituents is adopted. If the high priority groups lie of the same side, the configuration assigned is Z and if they are across it is *E*.

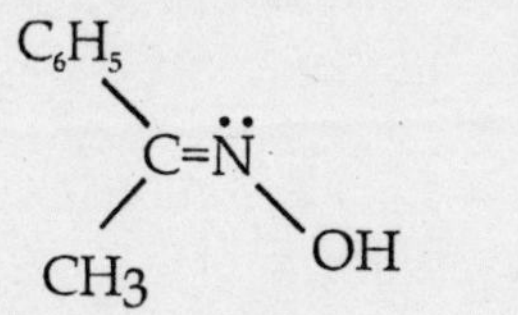

E-acetophenone oxime

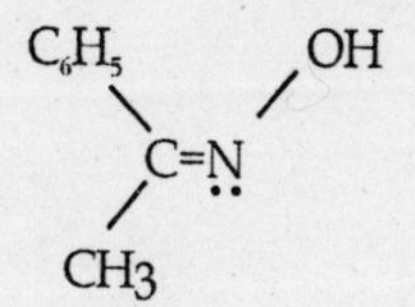

Z-acetophenone oxime

Z, *E*-system of nomenclature also applies to aldoximes:

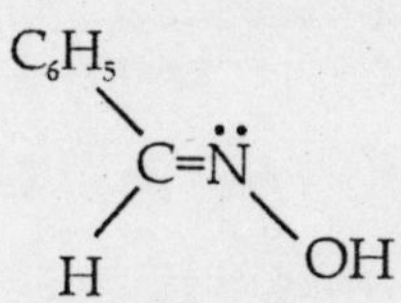

E-benzaldoxime

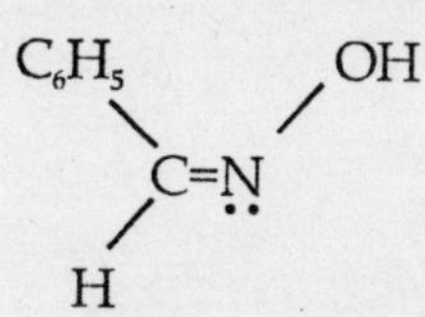

Z-benzaldoxime

Azobenzene which have nitrogen-nitrogen double bond exists in two geometrical isomeric forms *syn* and anti as shown below:

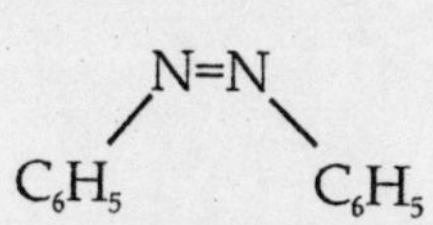

Syn-form

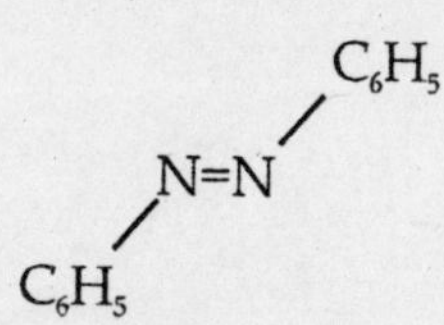

Anti form

Similarly diazoates also exist in two forms:

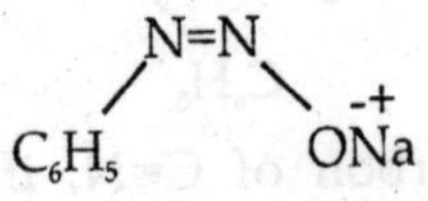

Syn-form

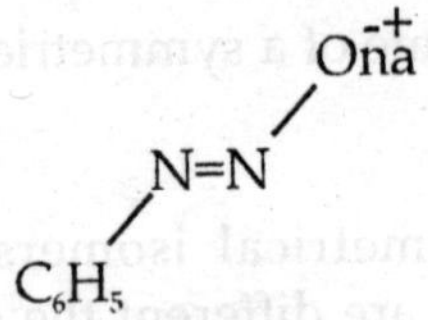

Anti form

Chapter 7

REACTION MECHANISMS-I

Organic reactions are quite a large in number and the compounds formed are numerous. Nevertheless, we can classify common organic reactions in the following five broad categories:

1 Substitution reactions

2 Addition reactions

3. Elimination reactions

4. Condensation reactions

5. Molecular rearrangements.

Our attempt here would be to see that mechanisms of these reactions mostly fit into a few basically simple patterns. Starting with the basic concepts, in this chapter we have utilized the basic principles in explaining interaction of molecules in various reactions.

BOND FISSION

An organic reaction involves (i)breaking of the bonds in reactant molecules releasing various short-lived reaction intermediates and (ii)formation of new bonds giving rise to the product molecules. Bond -fission or breaking may either be homolytic or heterolytic that depends on the nature of the reactants and condition of the reaction.

Homolytic fission—It involves a symmetrical breaking of the bond existing between the two atoms of a molecule (say, A B) releasing the groups (A and B), each of which retains one electron of the bonding electron pair—

$$A : B \longrightarrow A. + B.$$

Such groups with one unpaired electron are called free radicals. Reactions involving homolytic fission are called **free radical reactions.**

Heterolytic fission—It involves breaking of the bond existing between the two atoms of a molecule (say, A-A), releasing the groups (A and B) in such a way that one of the group retains the bonding electron pair. The group which retains the electron pair becomes negative-charged (anion) while the other becomes positive charged (cation)—

(i) $A/ : B \longrightarrow \overset{+}{A} + : \overset{-}{B}$

(ii) $A : /B \longrightarrow \overset{-}{A} : + \overset{+}{B}$

If the group has a negative charge on a carbon atom, it is called a carbanion and if it has a positive charge on a carbon atom, it is called a **carbonium ion.**

The reactions involving heterolytic bond-fission are known as ionic reactions and are initiated by following classes of reagent:

(i) Nucleophiles (nucleus-loving)—The reagents, which have an unshared pair of electrons readily available to an electron deficient atom for bond formation are called nucleophiles or nucleophilic reagents. The nucleophilic reagents may be neutral or may possess. a negative charge. A few common nucleophiles are

$$:\overset{-}{O}H, :\overset{-}{O}R, :\overset{-}{C}N, :\overset{-}{S}R\ RCO\overset{-}{O}:, :\overset{-}{Br}, :\overset{-}{I}, H_2\ddot{O}$$

$$R-\ddot{O}\text{-}H, R-\ddot{S}-H, \ddot{N}H_3, R_3\ddot{N}, R-\ddot{N}H_2 \text{ etc.}$$

(ii) Electrophiles (electron-loving) – The reagents which have an electron-deficient atom capable of bond-formation by readily accepting a pair of electrons, are called electrophiles or electrophilic reagents. A few common electrophiles are:

$$\overset{+}{H}_2,\ \overset{+}{N}_2O_2,\ SO_3,\ \overset{+}{Br},\ \overset{+}{Cl},\ \overset{+}{R},\ R\overset{+}{C}=O,\ BF_3,\ AlCl_3, \text{ etc.}$$

REACTION INTERMEDIATES

Highly reactive intermediates are formed in the course of organic reactions. The intermediates formed are temporary but play an important role in the process. Common of these are carbonium ions, carbanions, carbenes, nitrenes, free radicals and benzyne.

Carbonium ions

Cations having the positive charge on a carbon atom are called carbonium inons, *viz,* $-\overset{|}{\underset{|}{C}}{}^{+}$ A carbonium ion is formed when a group or atom leaves a carbon atom by taking away the bonding electron pair.

$$-\overset{|}{\underset{|}{C}}-X \longrightarrow -\overset{|}{\underset{|}{C}}{}^{+} + :\overset{-}{X}$$

A few carbonium ions are listed below –

$^{+}CH_3$ methyl carbonium ion	$+CH_2-CH_3$ ethyl carbonium ion	$^{+}CH_2-CH_2-CH_3$ *n*-propyl carbonium ion
$^{+}CH(CH_3)_2$ *iso*-propyl carbonium ion	$(CH_3)_3\overset{+}{C}$ *tert.* butyl carbonium ion	$CH_2{=}CH-\overset{+}{C}H_2$ *allyl* carbonium ion

$C_6H_5-\overset{+}{C}H_2$
benzyl
carbonium ion

Geometry of a carbonium ion—In a carbonium ion, the positively-charged carbon atom is sp^2- hybridised. So, a carbonium ion has a planar structure which contains three bonds of positively charged carbon atom are which 120° apart:

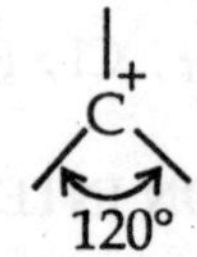

Stabilisation of carbonium icns—A carbonium ion is stabilised if its positive charge is dispersed from one carbon atom to other carbon atoms due to delocalisation of electrons. Dispersal of the charge occurs when the positively-charged corbon atom is (i) attached to double-bonded atoms, viz. $-\overset{|}{C}=\overset{|}{C}-\overset{+}{\underset{|}{C}}- \longleftrightarrow -\overset{+}{\underset{|}{C}}-\underset{|}{C}=\underset{|}{C}-$, or (ii) hyperconjugated with a σ (C − H) bond, viz., $-\overset{|}{\underset{H}{C}}-\overset{+}{\underset{|}{C}}- \longleftrightarrow -\overset{|}{\underset{H^+}{C}}=\underset{|}{C}-$.

Carbonium ions with these structural features are stabilised Actual structure of a carbonium ion,in such cases, is the resonance hybrid of the contributing structures. If the number of contributing structures is, more greater is the stabilisation of carbonium ions.

Dispersal of positive charge in allyl carbonium ion, benzyl carbonium ion and *tert.* butyl carbonium ion accounts for the stabilisation of these carbonium ions. Methyl carbonium ion ($^+CH_3$), on the other hand, is not stabilised because the positive charge in this case is localised on a single carbon atom.

$$CH_2 = CH - \overset{+}{C}H_2 \longleftrightarrow \overset{+}{C}H_2 - CH = CH_2$$

Resonating structures of allyl carbonium ion

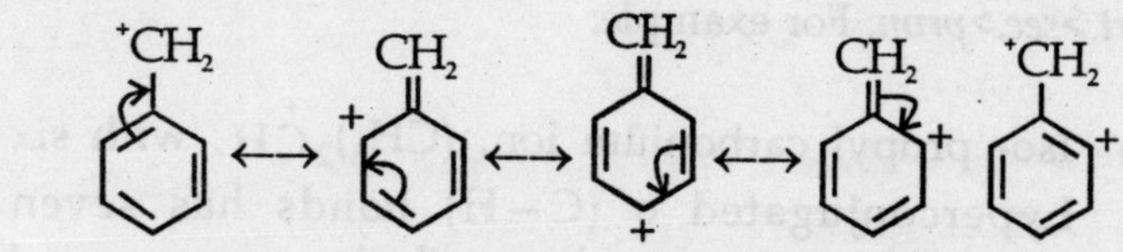

Resonating structures of benzyl carbonium ion

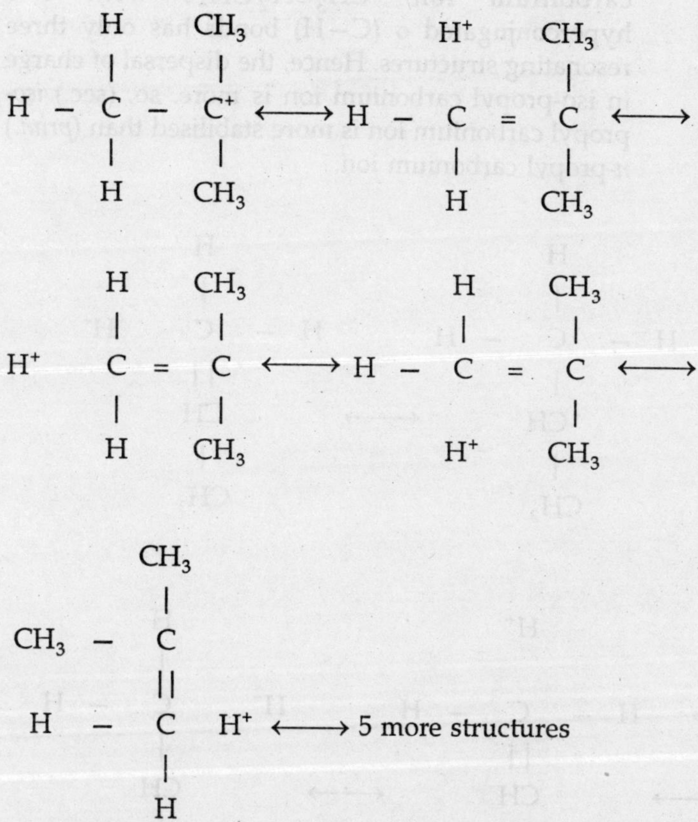

Resonating hyperconjugative structures of *tert*. butyl carbonium ion

Out of the two carbonium ions one with more number of hyperconjugated σ (C−H) bonds is more stabilised due to greater dispersal of charge. The relative extent of stabilisation of isomeric alkyl carbonium ios is thus in the order:

tert.>*sec.*>*prim*. For example:

(i) Iso- propyl carbonium ion, $(CH_3)_2\overset{+}{C}H$, with six hyperconjugated σ (C–H) bonds has seven resonating structures, whereas the isomeric n-propyl carbonium ion, $CH_3CH_2\overset{+}{C}H_2$, with two hyperconjugated σ (C–H) bonds has only three resonating structures. Hence, the dispersal of charge in iso-propyl carbonium ion is more. so, (sec.) *iso*-propyl carbonium ion is more stabilised than (*prim.*) *n*-propyl carbonium ion.

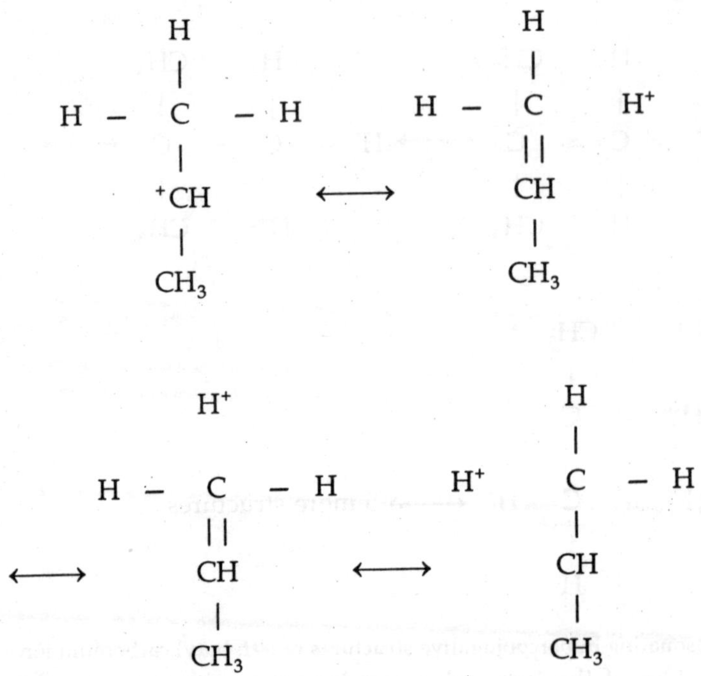

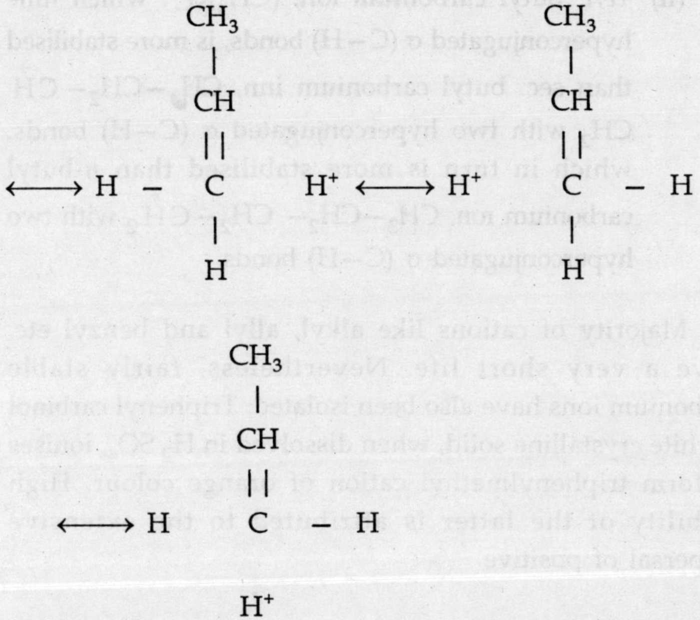

Resohating hyperconjugative structures of iso-propyl carbonium ion

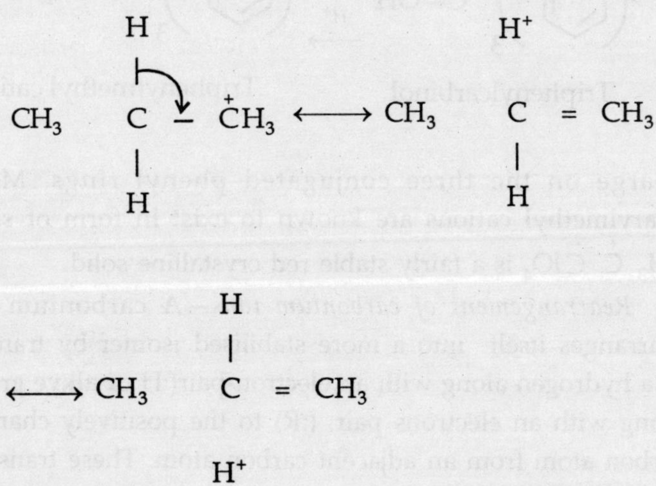

Resohating hyperconjugative structures of n- propyl carbonium ion .

(ii) *Tert.* butyl carbonium ion, $(CH_3)_3\overset{+}{C}$, which nine hyperconjugated σ (C–H) bonds, is more stabilised than sec. butyl carbonium inn, $CH_3-CH_2-\overset{+}{C}H$ CH_2 with two hyperconjugated σ (C–H) bonds, which in turn is more stabilised than *n*-butyl carbonium ion, $CH_3-CH_2-CH_2-\overset{+}{C}H_2$ with two hyperconjugated σ (C–H) bonds.

Majority of cations like alkyl, allyl and benzyl etc, have a very short life. Nevertheless, fairly stable carbonium ions have also been isolated. Triphenyl carbinol a white crystalline solid, when dissolved in H_2SO_4, ionises to form triphenylmethyl cation of orange colour. High stability of the latter is attributed to the extensive dispersal of positive

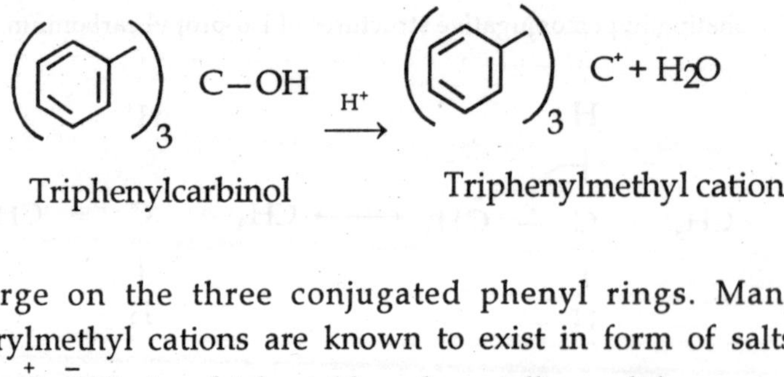

Triphenylcarbinol Triphenylmethyl cation

charge on the three conjugated phenyl rings. Many triarylmethyl cations are known to exist in form of salts. $PH_3\overset{+}{C}\overset{-}{C}lO_4$ is a fairly stable red crystalline solid.

Rearrangement of carbonium ions—A carbonium ion rearranges itself into a more stabilised isomer by transfer of a hydrogen along with an electron-pair(:H or alkye group along with an electrons pair. (:R) to the positively charged carbon atom from an adjacent carbon atom. These transfers are called *hydride-shift* and *alkyl-shift* respectively.

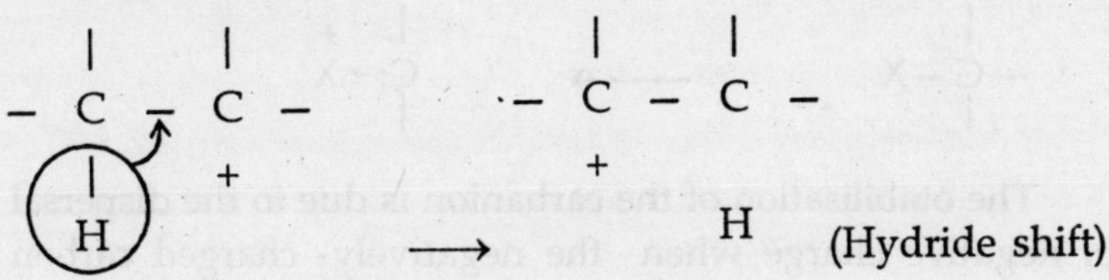

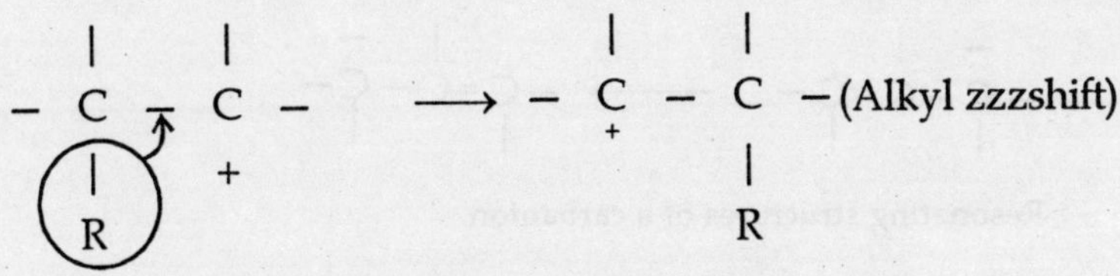

n-butyl carbonium ion rearrangement itself into more stabilised *sec.* butyl carbonium ion by hybrid shift:

$$CH_3-CH_2-CH_2-\overset{+}{C}H_2 \xrightarrow[\text{shift}]{\text{hydride–}} CH_3-CH_2-\overset{+}{C}H-CH_3$$

Rearrangement of *neo*-pentyl carboium ion into more stabilised 2-methyl-2-butyl carbonium ion involves alkyl shift:

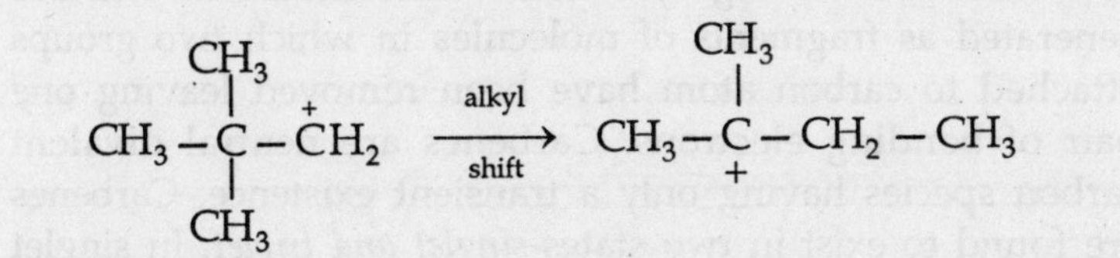

(neo-pentyl carbonium ion) (2-methyl-2-butyl carbonium ion)

Charbanions

Anions having the negative charge on a carbon atom are called carbanions, viz., $-\overset{..}{\underset{|}{C}}-$. This is the formation of a carbanion when a group or atom leaves a carbon atom without taking away the bonding electron-pair.

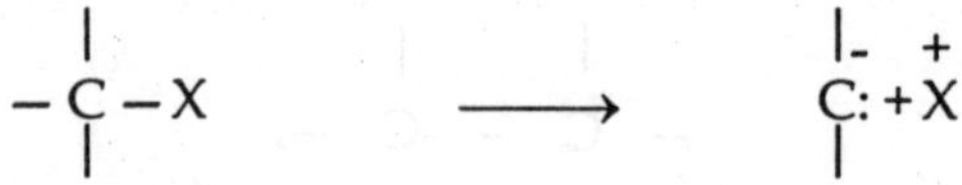

The otabilisation of the carbanion is due to the dispersal of negative charge when the negatively- charged carbon atom is attached with double-bonded atoms:

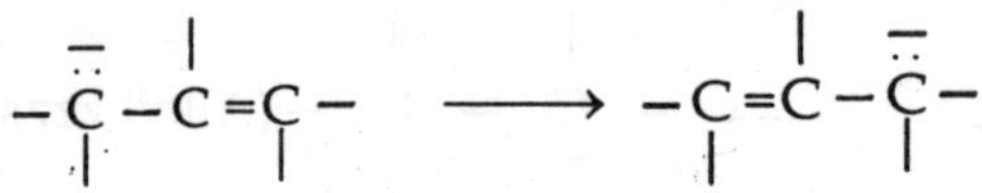

Resonating structures of a carbanion.

The structure of the group $-\overset{|}{\underset{|}{C}}-H$ in the molecule is of great importance. Its readiness to go over into a carbanion depends upon the ease with the negative charge is spread over rest of the molecule.

Carbenes

Carbenes are highly reactive intermediates and are generated as fragments of molecules in which two groups attached to carbon atom have been removed leaving one pair of bonding electrons. Carbenes are neutral divalent carbon species having only a transient existence. Carbenes are found to exist in two states-*singlet and triplet.* In singlet state, the carbon atom is sp^2 hybridised and the two unshared electrons are paired and have opposite spins, while in triplet, the carbon atom is sp hybridised and the two electrons are unpaired and have parallel spins. Methylene $(:CH_2)$, dichlorocarbene $(:CCl_2)$ benzylidene

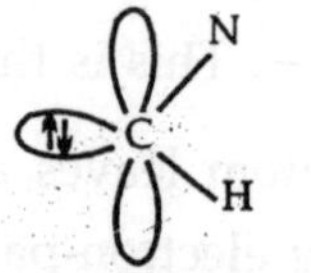

Singlet state of methylene

Triplet state of methylene

($:CH-C_6H_5$) and vinyl carbene ($:CH-CH=OH_2$) are some of the common carbenes.

Nitrens

Nitrenes ($R-\ddot{\underset{..}{N}}$, where R=halogen, alkyl, aryl, acyl, sulphonyl or amino etc.) have a monovalent electron-deficient nitrogen atom with only a sextet of electrons in its outermost shell. Nitrenes are iso- electronic with carbenes, also nitrenes exists in two states singlet and triplet. In the singlet state, there are two sets of paired electrons in sp^2 hybrid orbitals of nitrogen, whereas in the triplet, the nitrogen atom is sp hybridised and there are two free unpaired electron in two p orpbitals with parallel spins, one set of paired electrons residing in the sp hybrid orbital.

Singlet state of nitrenes   Triplet state of nitrenes

Benzyne

When aromatic halides are treated with certain strong bases, like KNH_2 undergo substitution of halogen by NH_2 not only directly but also at a position ortho to that formerly occupied by the halogen atom. In an experiment, labelled $1-{}^{14}C$ –chloroben-zene produced a mixture $1-{}^{14}C-$ and $2-{}^{14}C$ –aniline in the ratio of 48 : 52. Reactions of this type are called *cine-substitutions*

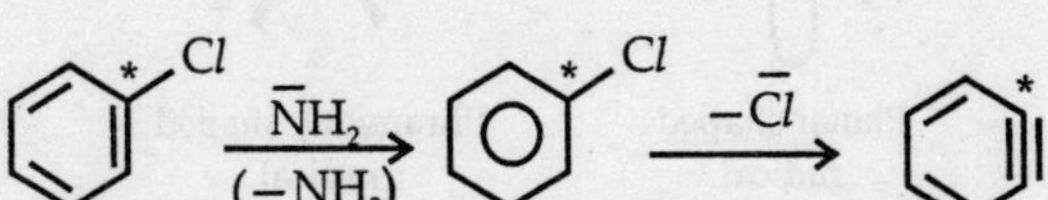

$1-{}^{*}C-$chlorobenzene   Benzyne

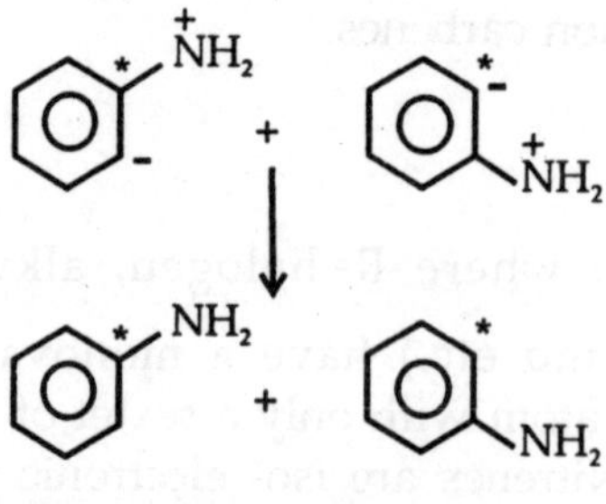

1–*C–aniline 2–*C–aniline

and require the intermediate formation of benzyne. Highly reactive and also a short lived intermediate. Following are the struetures: representing Benzyne:–

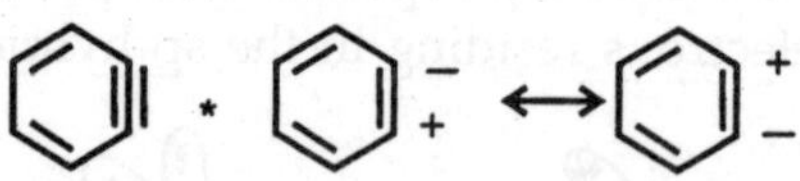

Free radicals

Intermediates having an unpaired electron on one of its atoms are called free radicals (or only radical) A free radical is formed when a group or an atom along with one electron of the bonding electron-pair is detached from a molecule:

$$-\overset{|}{\underset{|}{C}}-X \longrightarrow -\overset{|}{\underset{|}{C}}\cdot + \cdot X$$

Free radical

Radicals have been shown to exist in two shapes viz. planar and pyramidal.

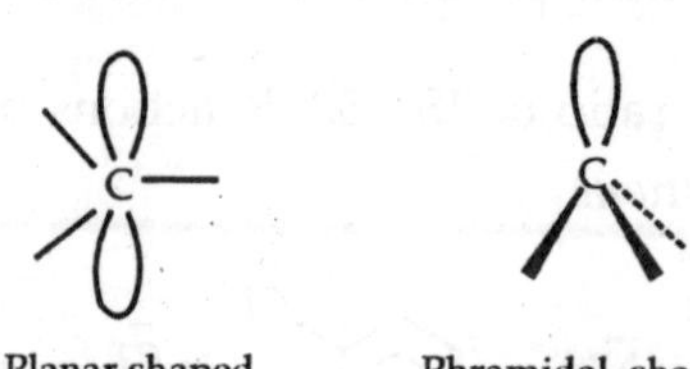

Planar shaped radical   Phramidal shaped radical

Free radicals are electrically neutral. They have a short life and are highly reactive. Once liberated, a free radical propagates a chain of reactions until either the reactant molecules are consumed or the reaction is terminated by inhibitors like hydroquinone or diphenylamine.

Certain compounds like 0–0, C–N, C–S, C–H and C–C contain weak bonds which form free radicals when decomposed thermally. For example, tetramethyl lead vapours on heating at 600–800° produce methyl free radicals.

$$Pb(CH_3)_4 \xrightarrow{600-800^\circ} Pb + 4.CH_3$$

Photo chemical reactions also produce free radicals. For example, chlorine molecule breaks into chlorine atoms in sunlight:

$$Cl-Cl \xrightarrow{hv} 2Cl.$$

Presence of reagents like organic peroxides, called initiators, induce the formation of free radicals.

$$RCO-O-O-CO\,R \xrightarrow{hv} 2\,R-CO-O$$

$$R-CO-O. \longrightarrow R. + CO_2$$

$$R. + HBr \longrightarrow R-H + .Br$$

Free radical reactions occur in gaseous phase or in nonpolar solvents. Such reactions follow the following coures:

(a) *Chain initiation*: The chain of radical reactions starts by the generation of a free radical as a result of homolytic cleavage of a covalent bond.

(b) *Chain propagation*: As soon as radicals, are formed they may undergo coupling, disproportionation, addition, substitution or rearrangement reaction.

$$R. + R. \longrightarrow R - R \qquad \textit{(coupling)}$$

$$CH_3CH_2{+}CH_3CH_2 \longrightarrow CH_2{=}CH_2{+}CH_3CH_3 \qquad \textit{(disproportionation)}$$

$$CH_2{=}CH_2{+}.Br \longrightarrow \underset{\displaystyle Br}{CH_2}-CH_2 \qquad \textit{(addition)}$$

$Br. + R-H \longrightarrow R. + HBr$ (*substitution*)

$Me_3CCH_2 \longrightarrow Me_2\,C-CH_2Me$ (*rearrangement*)

Stabilisation of free radicals—A free radical is stabilised if its unpaired electron is delocalised on several atoms. Delocalisation of the unpaired electron occurs if the atom carrying this electron is (*i*) attached to double-bonded atoms,

viz, 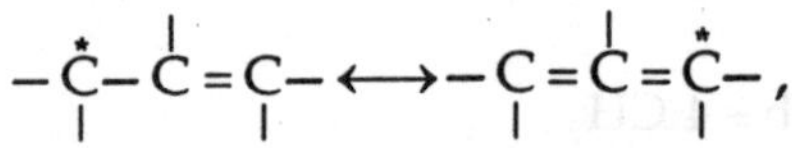 , or (*ii*) hyperconjugated

with a σ (C—H) bond, *viz.,* 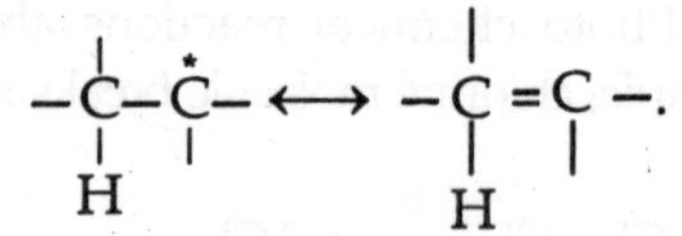. The actual structure of a free radical, which is more stabilised than any of the contributing structures, is a hybrid of all contributing structures. If the contributing structures are more in number higher is the stabilisation of free radical.

Triphenylmethyl radical is a fairly stable free radical, which can be stored. Its stability is explained by its existence as a resonance, hybrid of ten contributing structures:

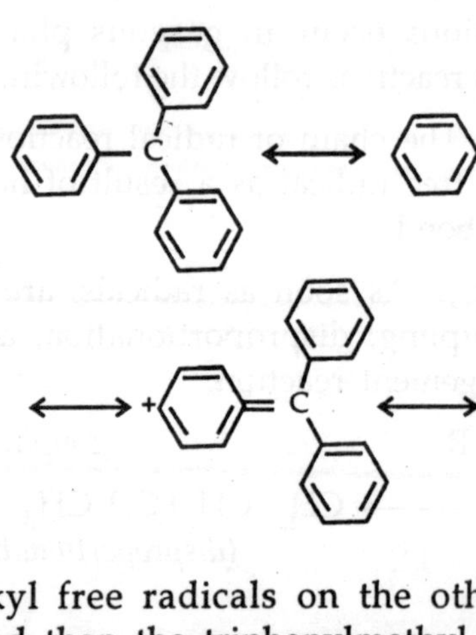

7 more structures

Alkyl free radicals on the other hand are far less stabilised than the triphenylmethyl free radical. To some extent they are stabilised by hyper conjugation. The order of stabilisation of the three types of alkly free radicals is :

tert.>*sec.*> *prim.* depending upon the number of hyperconjugated σ (C–H) bonds.

Thus, *tert* butyl free radical, $(CH_3)_3$ C, with nine hyperconjuga ted σ (C–H) bonds having ten resonating structures is more stabilised than sec. butyl free radical, CH_3, CH_2 CH CH.

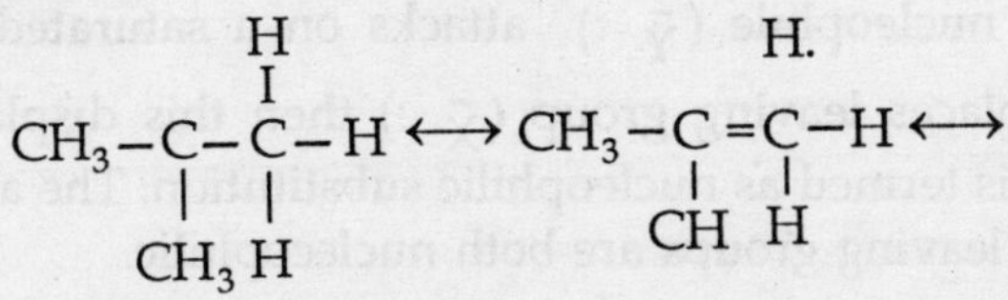

eight similar strutures resulting from remaining eight hyperconjugated (C–H) bonds.

(Resonating hyperconjgative structures of *tert.* butyl free radical)

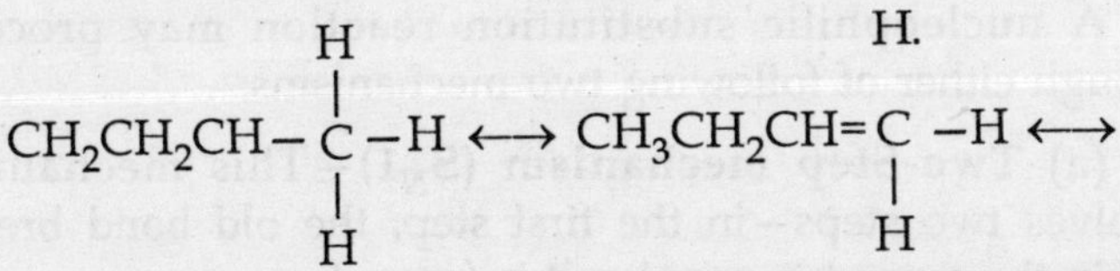

four similar structures result from remaining four hypercongated σ (c–bonds.)

(Resonating hyperconjgative structures of *sec.* butyl free radical)

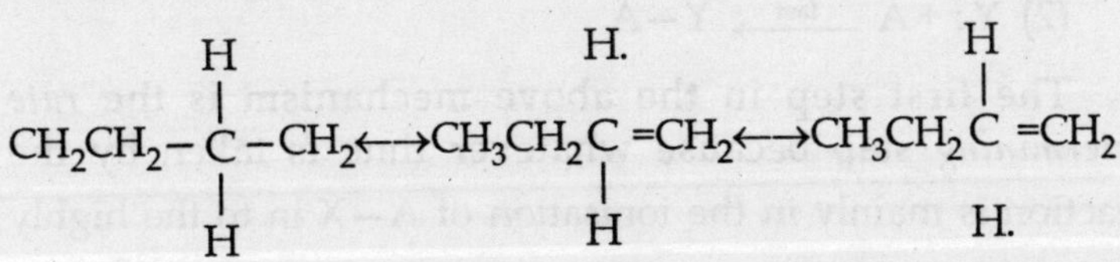

(Resonating hyperconjgative structures of *n*-butyl free radical)

with five hyperconjugated σ (C–H) bonds having six resonating structures, which in turn is more stabilised than *n*-butyl free radical with only two hyper conjugated σ (C–H) bonds having three resonating structures.

Isopropyl free radical, $(CH_3)_2$ HC is similarly more stabilised than *n*-propyl free radical, $CH_3-CH_2-CH_2$.

The relative ease of formation of alkyl free radical as intermediates in organic reactions depends upon their relative stabilisation.

SUBSTITUTION REACTIONS

(i) Nucleophilic substitution

If a nucleophile ($\bar{Y}$:) attacks on a saturated carbon and displaces leaving group ($\bar{X}$:) then this displacement reaction is termed as nucleophilic substitution. The attacking and the leaving groups are both nuclecophilic.

$$\bar{Y} : + A - X \longrightarrow Y - A + :\bar{X}$$

e.g., $:\bar{O}H + CH_3Cl \longrightarrow HO-CH_3 + C\bar{l}:$

A nucleophilic substitution reaction may proceed through either of following two mechanisms:

(a) **Two-Step mechanism (S_N1)**—This mechanism involves two steps—in the first step, the old bond breaks and in the second, a new bond is formed.

$$(1)\ A - X \xrightarrow{\text{slow}} \overset{+}{A} + \bar{X}$$

$$(2)\ \bar{Y} : + \overset{+}{A} \xrightarrow{\text{fast}} Y - A$$

The first step in the above mechanism is the *rate determining* step because whatever time is taken by the reaction is mainly in the ionisation of A—X in to the highly reactive carbonium ion $\overset{+}{A}$ and the leaving group $:\bar{X}$. No sooner $\overset{+}{A}$ is generated, it is attacked by the nucleophile $\bar{Y}$: to form the new bond (*In many cases ,the nucleophilic solvent molecules like* $H-\ddot{O}-H$ *and* $R-\ddot{O}-H$ *attack the carbonium ion* A^+ *and then the substitution is called solvolysis*). As the first step which is rate determining step does not involve the attacking nucleophile, and molar concentration of only one reactant (A−X)is changed, the two- step mechanism is of first order and is called S_N1, which means *substitution, nuleophilic, first order (or unimolecular).*

(b) **One step concerted mechanism (S_N2)**−In this mechanism, bond-making and bond-breaking are simultaneous or *concerted.*

$$\bar{Y}: + A-X \longrightarrow \left[\overset{\delta-}{Y}....A....\overset{\delta-}{X}\right]_{\text{transition state}} \longrightarrow Y-A+\bar{X}:$$

Here, $\bar{Y}$: attacks at the carbon in substrate A−X to form a high energy transition state(in which , bond Y−C is in the process of forming and C−X is in the process of breaking) with simultaneous secession of the leaving group :X. The molar concentration of two reactants is changed in this one step so its a bimolecular or second order reaction and is termed as S_N2, meaning *substitution, nucleophilic second order.*

The speed of ionisation step in S_N1 reactions depends upon the subsequent stabilisation of the carbonium ion and also upon the ionising power of the solvent. The rate is, however, independent of the concentration of the attacking nucleophile. Greater the stabilisation of carbonium ion, more is SN1 path favoured. The reactivity of carbon site is thus in the order: *tert.* >> *sec.* > *prim* > CH_3. On the other hand, the speed of S_N2 reactions is dependent on the concentration of nucleophile. The more important fact in S_N2 reaction is the finding of its way by attacking nucleophile to reach the carbon site and push away the leaving group. Crowding at the carbon site thus discourages such attack and the reactivity of the site is therefore in the order: CH_3 > *prim.* > *sec* >> *tert.*

Stereochemistry−Let us consider an optically active substrate R R' R'' $\overset{*}{C}$−X. As it is ionised , the sp^3 geometry of central carbon atom changes into sp^2 planar (symmetrical). Now, the attacking nucleophile $\bar{Y}$: has an equal opportunity to attack from either side of the planar carbonium ion with

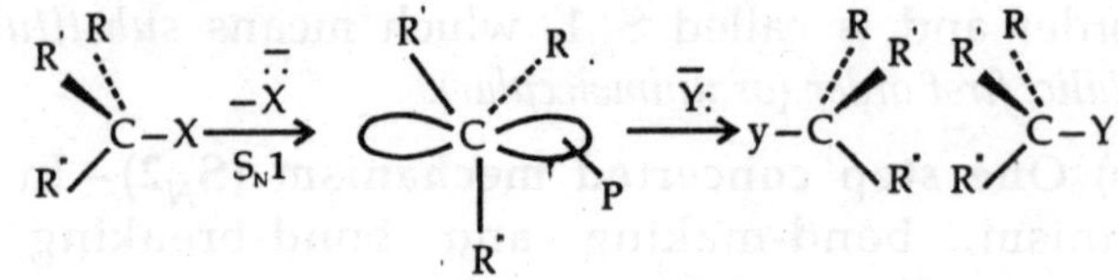

Optically active   Planar carbonium ion   Inversion of configuration   Retention of configuration

Racemization

equal possibility of inversion and retention of the original configuration. So, the final result is largely the racemization. stability of carbonium ion decides the extent of racemization. Greater the stability more is the racemization.

In S_{NZ} reactions, as the nucleophile $\bar{Y}$: approaches the carbon site of the substrate, leaving group starting detatching itself with the formation of high energy transition state. In the transition state, Y–C–X is

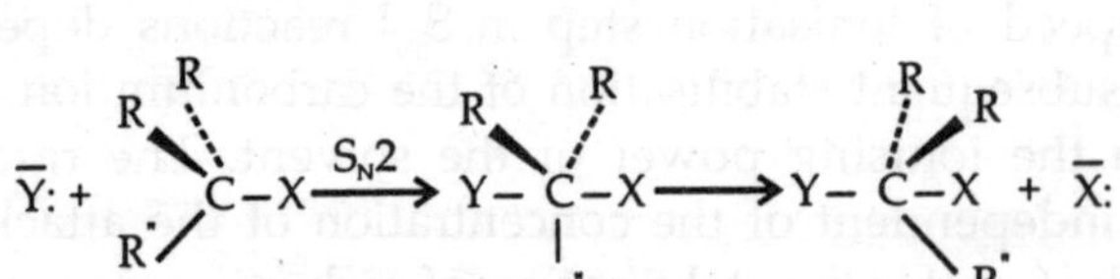

Optically active   Transition state   Inversion of configuration

perpendicular to the plane of C, R, R' and R". Now, as

soon as the leaving group :$\bar{X}$ is pushed away, the central-carbon is reversed like an inverted umbrella, the net result being the inversion of original configuration.

Effect of solvent: Ionising (or polar) solvents favour the reactions in which ions are produced, while non-polar solvents favour the reactions in which ions combine to form neutral molecules. Polarity (ionisig power) of some of the common solvents in is the order:

H_2O > HCOOH > Dimethylsulfoxide (DMOS) >

Dimethylformamide (DMF) > MeOH > ROH > Acetone > diethylether, Chloroform, Hydrocarbons.

Polar solvents mainly favour SN1 reactions but in case of SN2 ,Polarity is not a major criteria. Nevertheless, hydrogen-bonding solvents interact with nucleophiles and reduce their activity thus disfavouring SN2 path.

Steric effect—Another that plays an important role in deciding the mechanism of nucleophilic substitutions is the *extent of mechanical interference of the groups* at the carbon site. This is known as Steric effect (effect of space). The space around carbon atom becomes more and more crowded with increase in the number and size of the alkyl groups. Under such a condition, formation of transition state in SN2 mechanism becomes increasingly difficult due to lack of space at this carbon atom. This is known as steric *hindrance* for SN2 mechanism. On the other hand, crowding increases the strain (Steric strain) upon C—X bond, thereby increasing the ionisation. This is called *steric acceleration* for SN1 mechanism. For example, in tertiary alkyl halides (R_3C-X) with maximum crowding at α carbon atom, there is so much steric hindrance for SN2 and steric acceleration for SN1 mechanism that the hydrolysis of these halides proceeds entirely through SN1 mechanism. The steric effect in primary and secondary alkyl halides, is not so much noticeable and therefore the reaction proceeds through both SN1 and SN2 mechanisms.

When the reaction prefers SN1 mechanism due to steric effect, the carbonium ion produced in its ionisation step, is expected to combine with the nucleophile in the second step more slowly, because in doing so, the steric strain will be reintroduced into the molecule. In such cases, the carbonium ion tends to undergo elimination, resulting into the formation of an olefin. This is illustrated below in the case of tertiary butyl bromide—

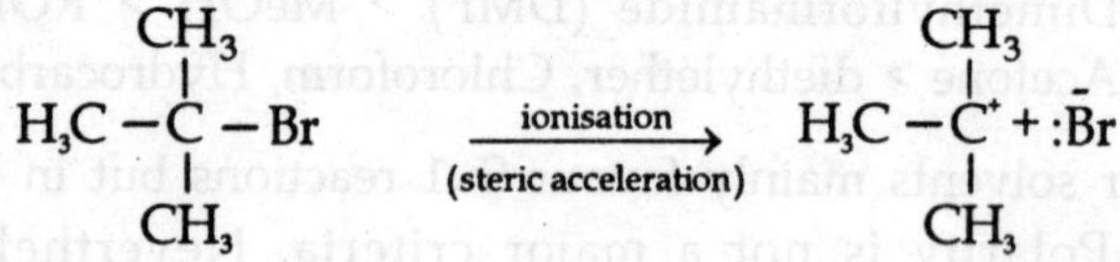

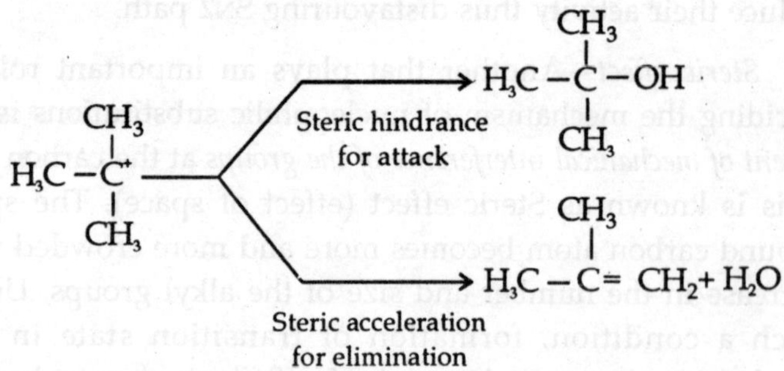

(ii) Electrophilic substitution

The conjugated π electron system of an aromatic ring is the source of electrons for an electrophile. Activating groups like OH, NH_2, OR etc., if already attached to the ring, further increase the electron density of the ring due to resonance. Electrophilic substitution may be defined as the substition of hydrogen on the aromatic ring by an electrophile. For example,

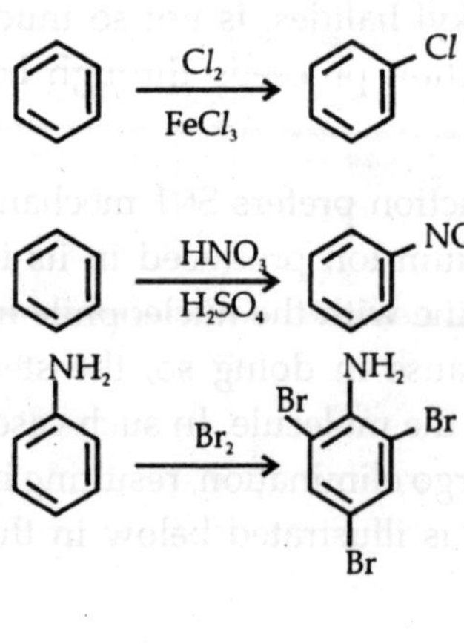

The general mechanism applicable to all aromatic electropnilic substitutions is described below showing the attack of electrophile E^+ on a benzene ring.

Generation of the electrophile E^+

$Nu-E \rightleftharpoons N\bar{u}: + E^+$

Attack of E^+

E^+ attacks the π electron-pair of the benzene ring to form a carbonium ion as intermediate. This cation is stabilised by resonance and further addition of nucleophile on this system is not possible (*cf.* alkenes undergo such addition)

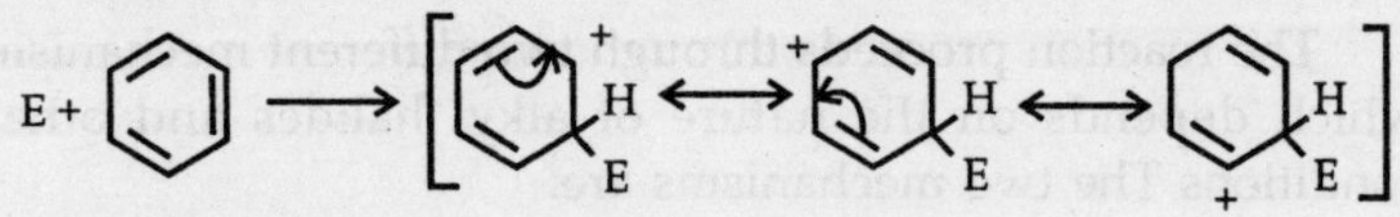

Aromatisatio

A proton is then abstracted by $N\bar{u}:$ from the carbonium ion to regenerate a stable aromatic system.

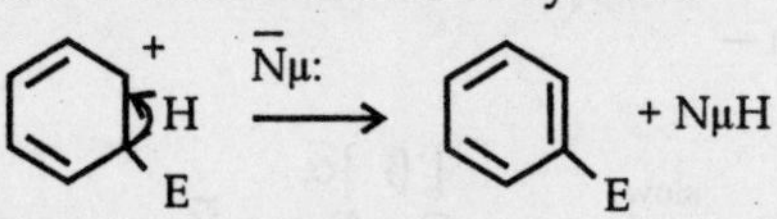

(iii) Substitution of Free radical

Free radical substitution adopts a chain reaction course involving three steps, viz, initiation, propagation and termination. For example chlorination of methane, in sunlight is a free radical subsitution. It proceeds as below:

$Cl_2 \xrightarrow{hv} Cl. + \cdot Cl$ (*initiation*)

$CH4 + \cdot Cl \longrightarrow \cdot CH_3 + HCl$ (*propagation*)

$Cl_2 + \cdot CH_3 \longrightarrow CH_3Cl + \cdot Cl$

$Cl. + Cl. \longrightarrow Cl_2$ (*termination*)

ELIMINATION REACTIONS

In these type of reactions, two groups or atoms attached to two adjacent carbon atoms in a molecule are lost with simultaneous formation of a multiple bond (double or triple) between the two carbon atoms. In elimination reaction there is loss of a nucleophile from one carbon atom and a proton from the adjacent one. A typical example of such reactions is the formation of alkene by dehydrohalogenation of alkyl halide in presence of a strong base—

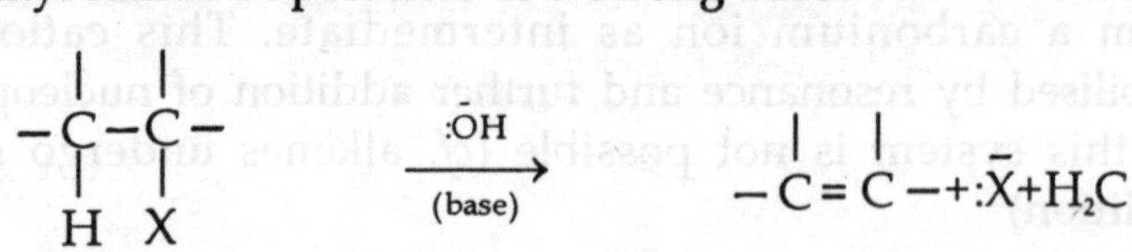

The reaction proceeds through two different mechanism which depends on the nature of alkyl halides and other conditions The two mechanisms are:

(a) Two-step reaction mechanism(E_1)

First-step: Ionisation of the alkyl halide to release the nucleophile $:\bar{X}$ from the α carbon atom, producing alkyl carbonium ion—

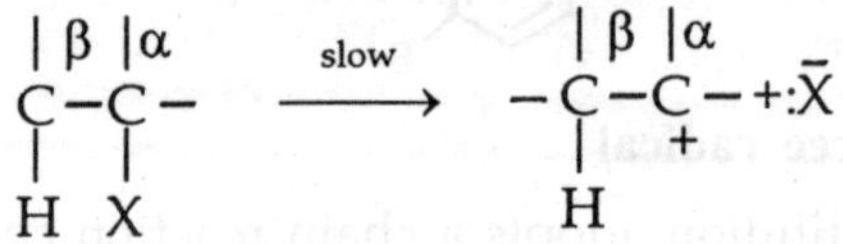

Second step: Elimination of proton from β carbon atom to produce the alkene—

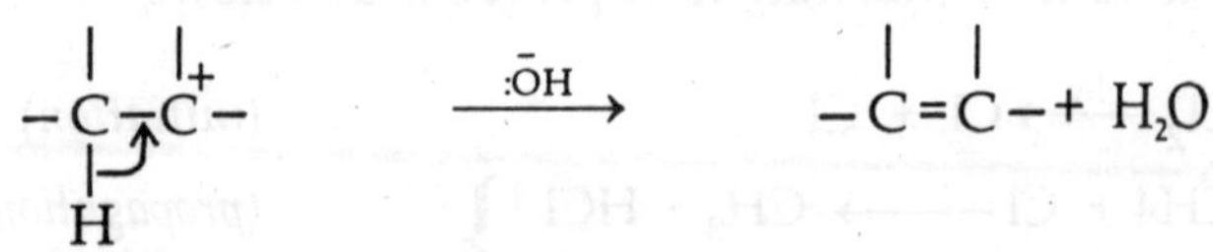

The rate of the overall reaction is determined by slow ionisation of C—X bond to form the carbonium ion; because once the carbonium ion is formed, it reacts rapidly to form the product. First step is thus the *rate determining step.* As in this mechanism there is change in molar concentration of one reactant in the rate-determining step, the reaction is *unimolecular or of first order*. Such a mechanism is termed E_1 (*elimination, first order*).

(b) One step reaction mechanism (E_2)

This mechanism involves the abstraction of a proton by base from β carbon atom and simultaneous release of nucleophile $:\bar{X}$ from the carbon atom of the alkyl halide molecule.

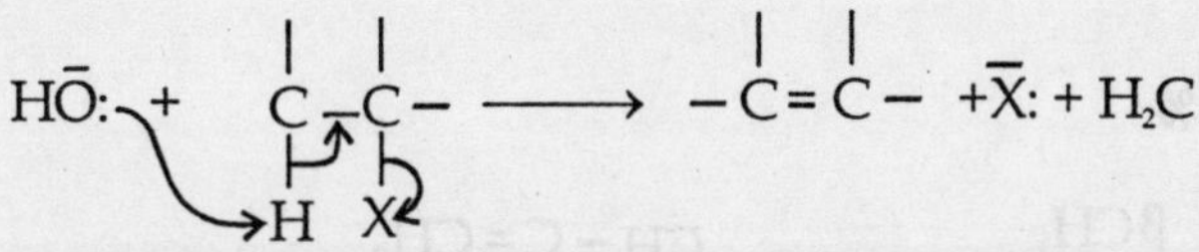

Since in this mechanism, the change *in molar concentration* of two reactants is involved, the reaction is *bimolecular* or of *second order*. Such mechanism is called E_2 (*elimination, second order*).

The elimination reactions and nucleophilic substitutions are closely related and competitive in nature. In presence of strong bases (alcoholic KOH, amines etc.) in the medium, alkyl halides prefer to undergo elimination reaction, whereas in presence of weak bases (AgOH, aqueous KOH, etc.) substitution is favoured. If the size and number of alkyl groups at the λ-carbon atom increase the tendency of alkyl halides to undergo elimination reaction increases and that for substitution decreases . This is illustrated in the following examples.

$$\overset{\beta}{C}H_3-\overset{\alpha}{C}H_2-Br \xrightarrow[\text{Alc.KOH}]{\bar{R}\bar{O}\overset{+}{K}} \underset{90\%}{CH_3-CH_2-OR}+\underset{91\%}{CH_2=CH_2}$$

$$CH_3-\overset{\beta}{C}H_2-\overset{\alpha}{C}H_2-Br \xrightarrow[\text{Alc.KOH}]{\bar{R}\bar{O}\overset{+}{K}} CH_3-\underset{64\%}{CH_2}-CH_2-$$

$OR + CH_3\,CH{=}CH_2$

36%

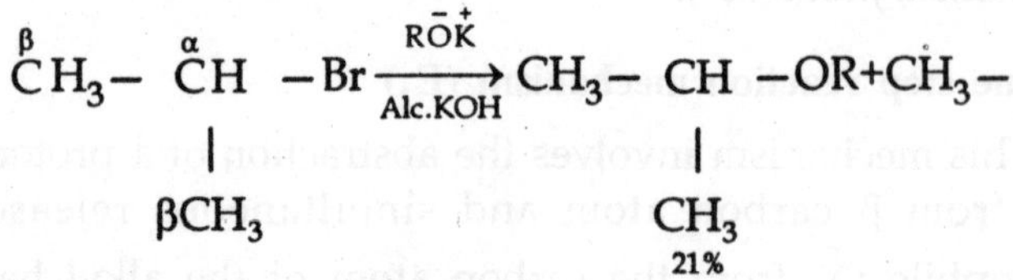

$CH{=}CH_2$

79%

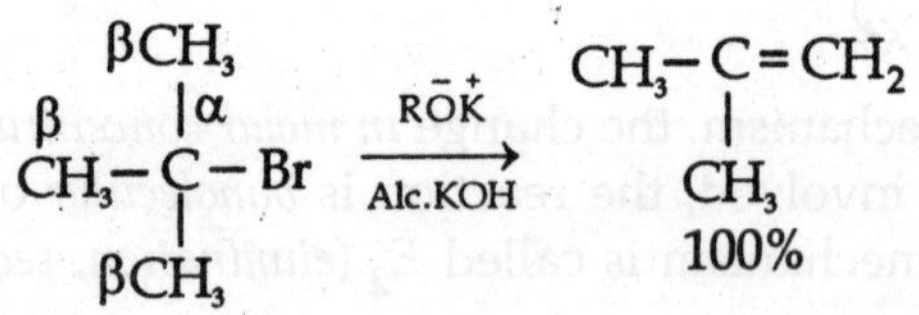

The conditions which influence the elimination to proceed through E_1 or E_2 mechanism are the same as those for S_N1 and and S_N2 respectively.

In dehydrohalogenation of alkyl halides having dissimilar β carbon atoms, proton is eliminated preferentially from the more substituted β carbon atom *i.e.* the one which is poorer in hydrogen (Saytzeff's rule). For example, dehydrohalogenation of 2-bromobutane, as illustrated in the reaction given below, can produce butene-1, as well as butene-2, but butene-2 is preferentially formed and is the main product.

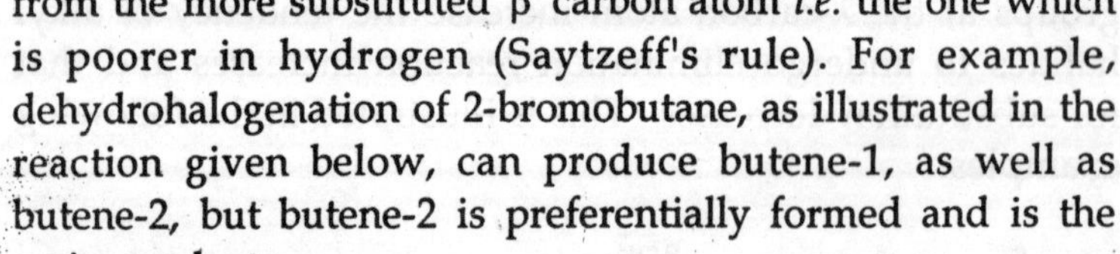

2–bromobutene

Using saytzeff's rule we can explain the preferential formation of product on the basis of hyper conjugation. Butene-2 is preferentially formed here because of its greater stabilisation by hyperconjugation than that of butene-1 (six C—H) bonds being hyperconjugated in butene-2, while only two in butene-1).

ADDITION REACTIONS

(i) Addition to alkenes

Alkenes form addition compounds with halogens, halogen acid, sulfuric acid, hypohalous acids and nitrosyl halides etc. These addition reactions proceed either by ionic or free radical mechanism. The mechanism of addition is ionic in the absence of sunlight or peroxides.

Ionic mechanism—This is a two step mechanism. In the first step, the adding molecule (X—Y), on coming close to the π electrons of the double bond undergoes polarisation $\left(\overset{\delta_+}{X}-\overset{\delta_-}{Y}\right)$. At the same time, the π electron pair undergoes electromeric displacement and migrates to one of the π bonded carbon atoms, from where it attacks the electrophilic end $\left(\overset{\delta_+}{X}\right)$ of the adding molecule. This attack brings about the union of the electrophilic end of the adding molecule with the olefinic carbon atom to which the electron pair had migrated, forming a carbonium ion (the product of electrophilic addition). Simultaneously, the nucleophilic portion ($\bar{Y}$:) of the adding molecule is released. In the second step, the nucleophile $\bar{Y}$: is bonded to the positive - charged carbon atom of the carbonium ion to produce the final addition compound—

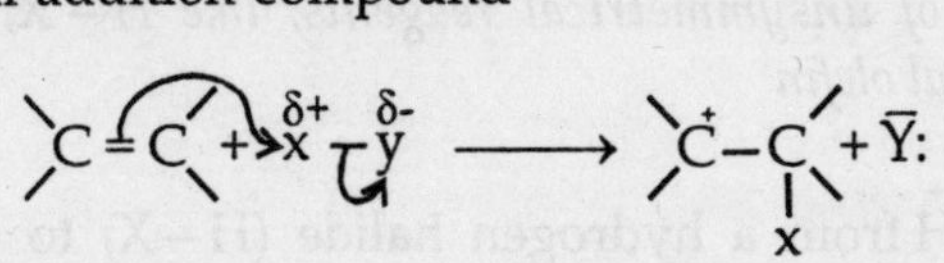

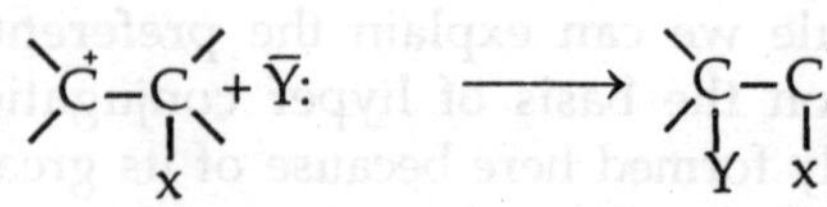

In the two step mechanism stated above, the electrophilic addition occurs first than nucleophilic addition. This is because of (*i*) inability of a nucleophile to attack a bonded system, the latter itself being a nucleophile and (*ii*) lack of stabilisation of the carbanion (if produced).

Depending on the nature of the adding molecule and the structure of olefin, the addition reactions proceeding by ionic mechanism are classified under the following three categories;

(*i*) *Addition of symmetrical reagents, like halogens*

In the addition of say, bromine to olefins the former is polarised ($\overset{\delta+}{Br} - \overset{\delta-}{Br}$) on its approach to π electron pair of the olefin. The π electron pair then attacks the positive end of the bromine molecule, uniting one of the olefinic carbon atoms, with this end and, simultaneously, releasing $:\bar{Br}$. The other olefinic carbon atom, thus, acquires a positive charge to form a carbonium ion. In the second step, the $:\bar{Br}$ anion released in the first step, combines with the carbonium ion forming the final addition product—

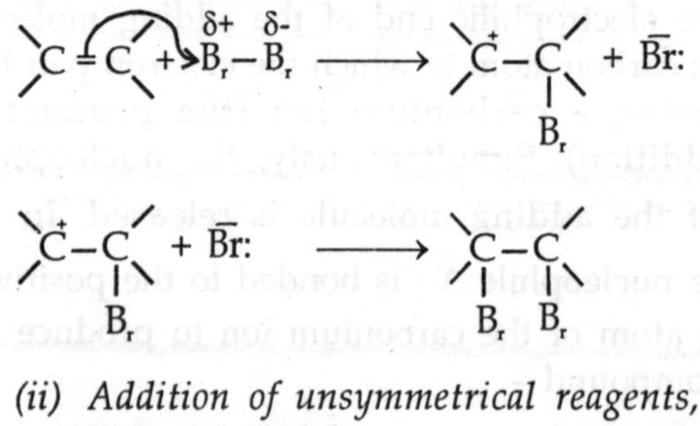

(*ii*) *Addition of unsymmetrical reagents, like H – X, in symmetrical olefin*

Addition of $\overset{+}{H}$ from a hydrogen halide (H–X) to the olefin produces the intermediate carbonium ion. The additon is then completed by combination of $:\bar{X}$ with the carbonium ion in second step—

(iii) Unsymmetrical Reagent Addition like H–X in unsymmetrical olefins

When the olefin is unsymmetrical, such as propene, addition of H–X can theoretically form two products, as shown below—

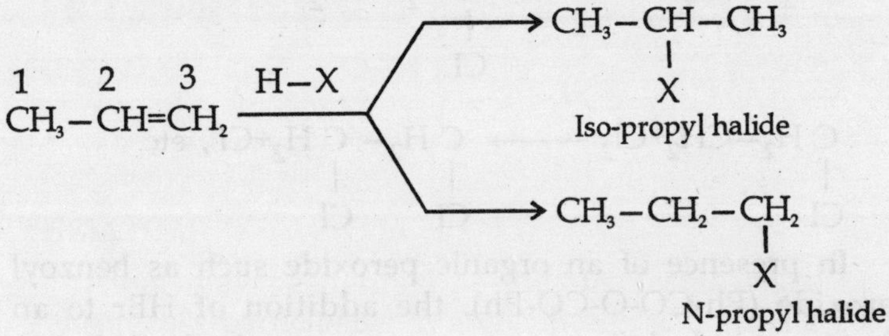

However, in actual reaction iso- propyl halide is the main product. This preferential formation of iso-propyl halide may be due to the +E effect of methyl group in propene. This effect tends to displace the π electron -pair on C_1 which then unites with H of H–X; and the nucleophile $:X^-$ unites with C_2. Another factor which is responsible for this course of addition may be the greater. stabilisation of *iso*-propyl (*secondary*) carbonium ion ($CH_3-\overset{+}{C}H-CH_3$) than of n-propyl (primary) carbonium ion ($CH_3-CH_2-\overset{+}{C}H_2$) due to hyperconjugation. This explains **Markownikof's rule** according to which, in addition of an unsymmetrical molecule to an unsymmetrical olefin, the negative part of the adding molecule is added to the more substituted olefinic

---

* The bond in H-Cl in stronger than that is H-Br and can not be broken homolytically. In H-I, the bond breaks as in H-Br, but Iodine is not reactive enough to be added to a double bond. The two Iodine atoms couple to make Iodine molecule. Hence, peroxide is specific to H-Br. (Bond energies of H-Cl, H- Br and H-I are 103.1, 87.4 and 71. kcal/ mole respectively]

carbon atom *i.e.* the one having lesser number of hydrogen atoms.

Free radical mechanism—The addition of halogen to alkenes in sunlight follows a free radical path:

$$Cl_2 \xrightarrow{h_2} Cl. + Cl.$$

$$CH_2 = CH_2 + Cl \longrightarrow \underset{\substack{| \\ Cl}}{C}H_2-CH_2$$

$$\underset{\substack{| \\ Cl}}{C}H_2-CH_2+Cl_2 \longrightarrow \underset{\substack{| \\ Cl}}{C}H_2-\underset{\substack{| \\ Cl}}{C}H_2+Cl., \text{ etc.}$$

In presence of an organic peroxide such as benzoyl peroxide (Ph-CO-O-CO-Ph), the addition of HBr to an unsymmetrical olefin, say propene, proceeds through free radical mechanism giving rise to l- bromopropane, a product opposite to that predicted by Markownikoff's rule. This is known as *kharasch peroxide* effect. HCl, HI and HF do not* exhibit this abnormal reaction. This peroxide effect is a free radical chain reaction that involves following steps:

Initiation:

$$\underset{\text{Organic peroxide}}{R-\overset{\overset{O}{||}}{C}-O-O-\overset{\overset{O}{||}}{C}-R} \xrightarrow[\text{light}]{hv} \underset{\text{Carboxylate free radical}}{2R-\overset{\overset{O}{||}}{C}-O}$$

$$R-\overset{\overset{O}{||}}{C}-O. \longrightarrow R. + CO_2$$

$$R. + H-Br \longrightarrow R-H + .Br$$

Propagation by addition—Free Bromine radical now attacks the π-electron of the olefin and may lead to the formation of two different types of free radicals:

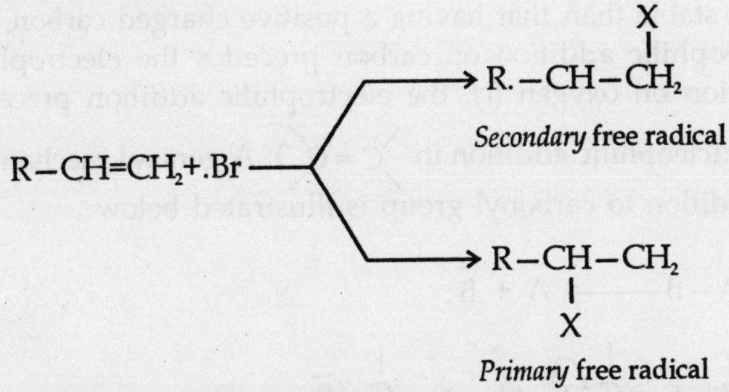

Since due to hyperconjugation, the *secondary* free radical is more stablised than the *primary* free radical, the former is preferentially formed in the above step of the addition reaction. Further propagation continues as show in the reaction below, finally producing the addition compound.

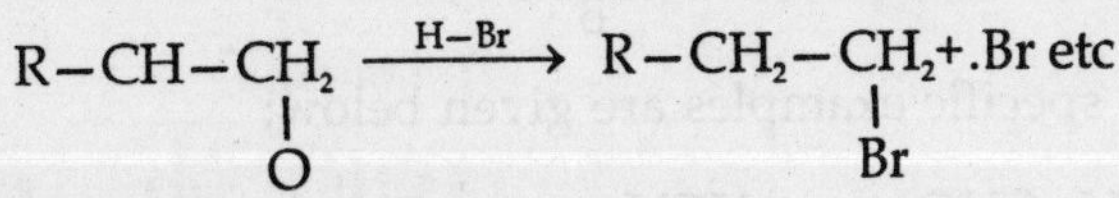

Termination:

$$Br. + .Br \xrightarrow{\text{Inhibitor}} Br-Br$$

(ii) Addition to carbonyl group

The carbonyl function owes its activity to the polarisation of π electrons between the carbon atom and the more electronegative oxygen due to mesomeric effect:

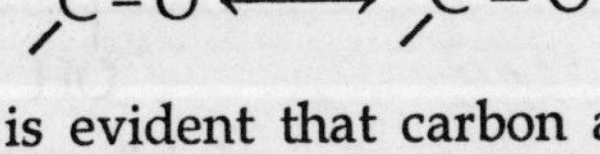

Hence it is evident that carbon atom carbonyl group acquires a positive charge which becomes a site for nucleophilic attack. By the attack of a nucleophile on the carbonyl carbon atom, the electron pair shifts over the oxygen atom making latter the centre of electrophilic attack. As a system which have negative charged oxygen atom is more stable than that having a positive charged carbon, the nucleophilic addition on carbon precedes the electrophilic addition on oxygen (*cf.* the electrophilic addition precedes the nucleophilic addition in $\rangle C=C \langle$). A general mechanism of addition to carbonyl group is illustrated below:

$$A-B \longrightarrow \overset{+}{A} + \overset{-}{B}:$$

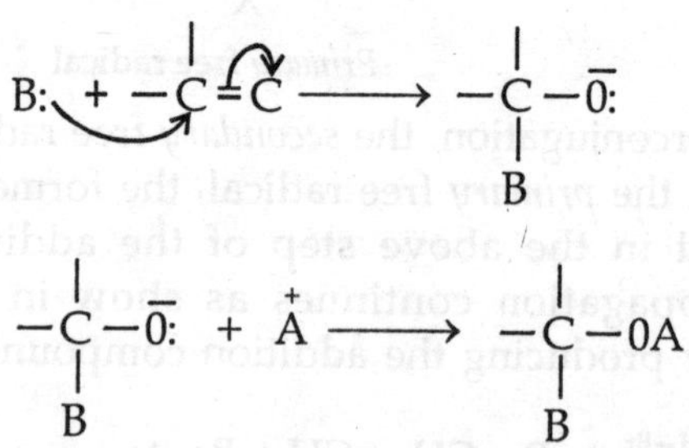

A few specific examples are given below:

1. $CH_3\,CHO + HCN \longrightarrow CH_3\,\underset{\substack{| \\ OH}}{CH}\,CN$

Cyanohydrin

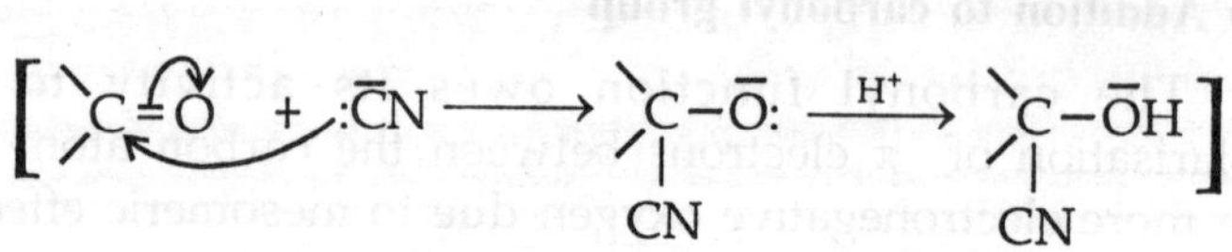

2. $CH_3\,CHO + Na\,HSO_3 \rightleftharpoons CH_3\underset{\substack{| \\ OH}}{CH}\overline{So}_3\overset{+}{Na}$

Sod. Bisulfite adduct

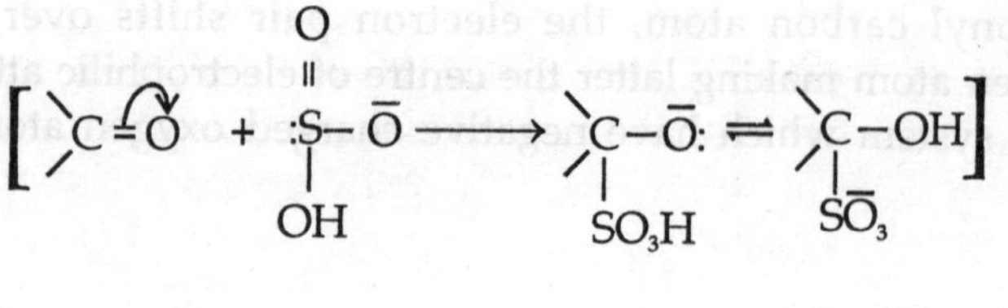

$$Ph\,CHO + RNH_2 \longrightarrow \left[\begin{array}{c} Ph\,CH\,NHR \\ | \\ OH \end{array}\right] \xrightarrow{-H_2O} Ph\,CH{=}N{-}R$$

Not isolable

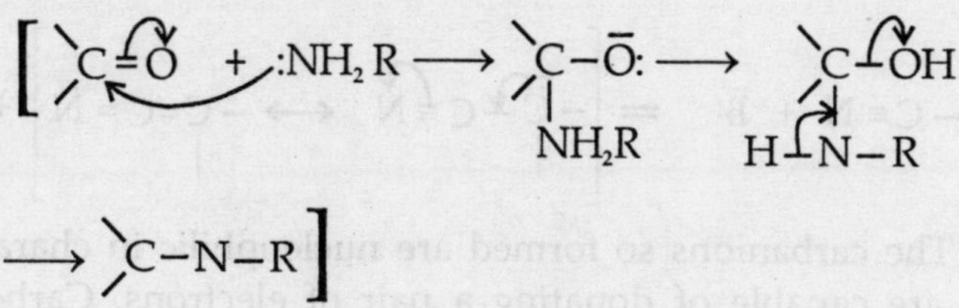

CONDENSATION REACTIONS

A condensation reaction may be defined as the reaction in which intermolecular elimination of a water molecule takes place from the two combining molecules forming new C−C bonds.

Resonance-stabilised carbanions are formed by abstraction of weakly acidic hydrogen from the compounds having electron withdrawing groups like carbonyl, nitro, ester and nitrile etc., by the action of a strong base such as NaOEt, $NaNH_2$, *t*-BuOK, NaH and triphenylmethyl sodium ($Ph_3\overset{-}{C}\overset{+}{Na}$) etc.

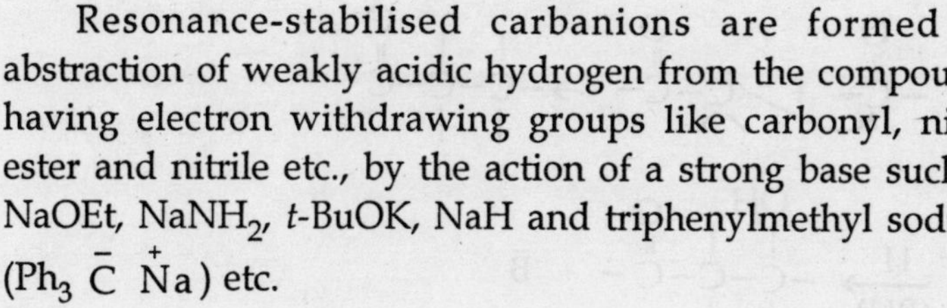

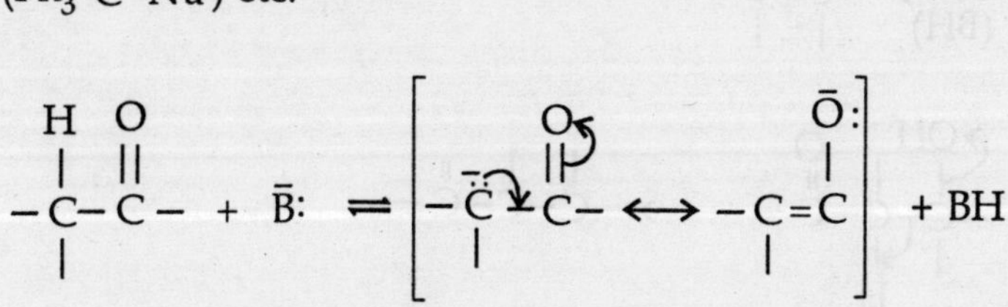

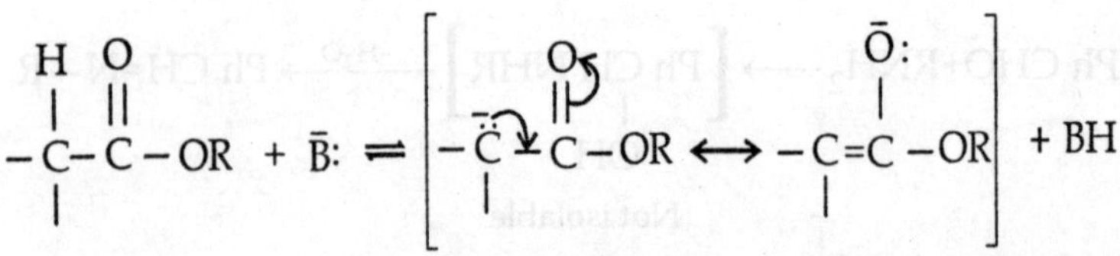

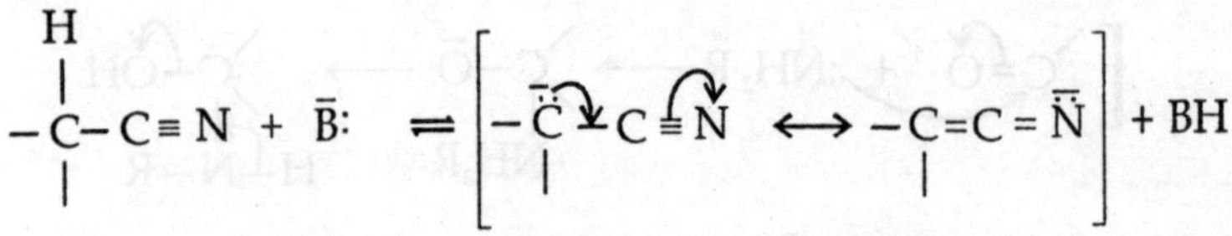

The carbanions so formed are nucleophilic in character and are capable of donating a pair of electrons. Carbonyl group itself is an electron-acceptor (as explained earlier) and hence is attacked at the carbon atom by the carbanion with the formation of a new C–C bond. The resulting anion gets protonated ($\overset{+}{H}$ is supplied by BH) to form a β hydroxy compound. The latter loses a molecule of water by simple elimination if an α hydrogen is present:

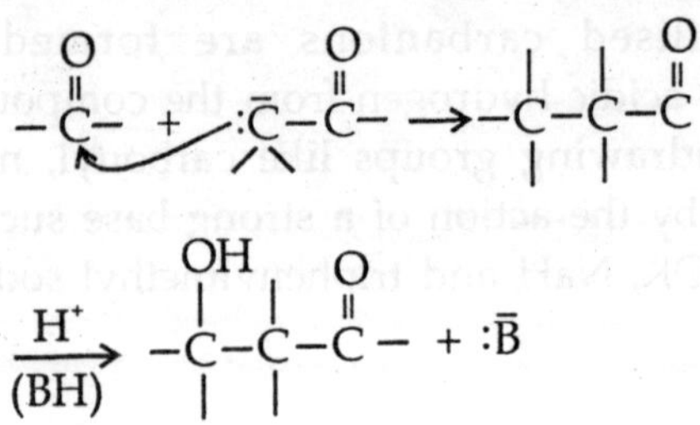

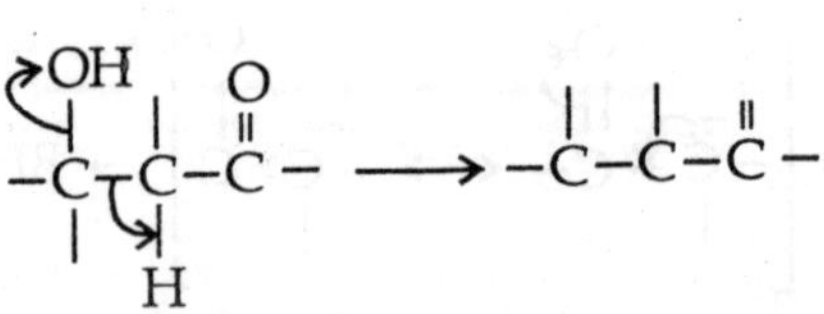

The mechanism described above is a general route followed by a variety of condensation reactions namely *Aldol condensation, Knoevenagel reaction, Claisen reaction* and *Perkin reaction* etc. covered in Chapter 8.

MOLECULAR REARRANGEMENTS

A group when migrates from one atom to another within the same molecule which changes the carbon skeleton to more stablised state, it is termed as *molecular rearrangements.*

(i) Migration to electron-deficient carbon

A carbonium ion formed as an intermediate in a reaction may rearrange into a more stabilised ion by *1,2 – shift* of a hydrogen atom, an alkyl or an aryl group.

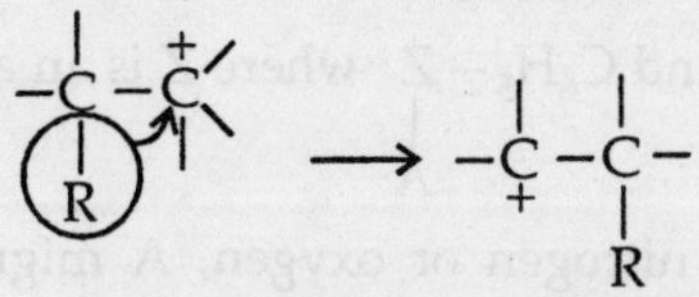

A few reactions in which such a rearrangement takes place are *Wagner-Meerwein* and *pinacol-pinacolone* rearrangenments (see chapter 8).

(ii) Migration to electron-deficient nitrogen

Nitrenes undergo a rearragement in which an alkyl, aryl or H migrates from α carbon to the electron-deficient nitrogen atom by 1, 2-shift.

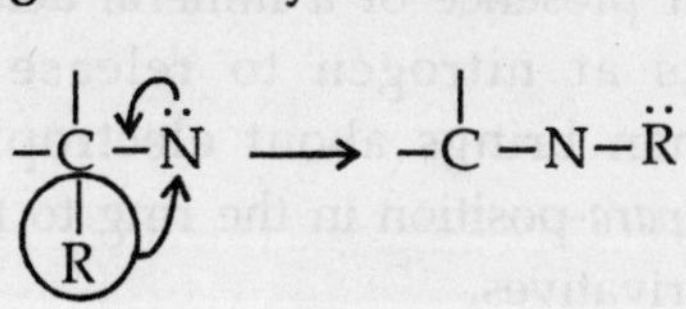

There are many reactions which involve such rearrangements like Hofmann degradation of amides, Curtius rearrangement and schmidt rearrangement are some of them described in chapter 8.

(iii) Migration to electron- deficient oxygen

An alkyl or phenyl group migrates from a carbon to the electron-deficient oxygen atom by 1, 2-shift.

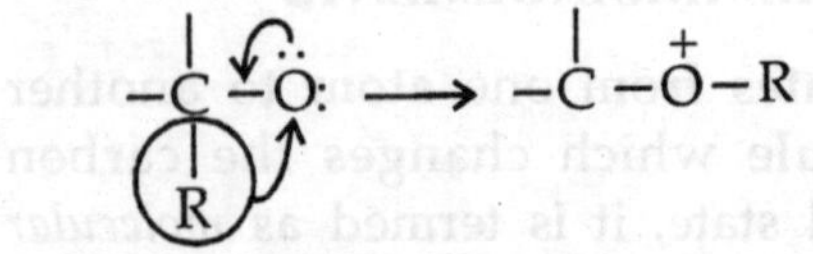

Baeyer-Villiger oxidation and rearrangement of *cumene. hydroperoxide* into *phenol* are few examples which involves migration of an alkyl or aryl group to the electron deficient oxygen (see chapter 8).

(iv) Rearrangement of groups in an aromatic ring

In an aromatic compound $C_6H_5-\underset{A}{\underset{|}{Z}}$ where Z is an atom other than carbon-usually nitrogen or oxygen, A migrates from Z to an *ortho*-or a *para position* in the nucleus. Such rearrangements are usuallly acid-catalysed.

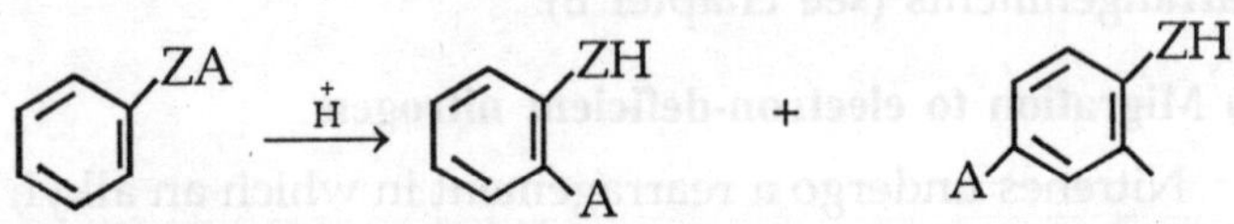

N-substituted aniline derivative(C_6H_5--N–A) undergo molecular rearrangement in presence of a mineral acid. A proton from acid attacks at nitrogen to release the electrophile A^+, which then brings about electrophilic substitution at the *ortho* or *para*-position in the ring to form ring- substituted aniline derivatives.

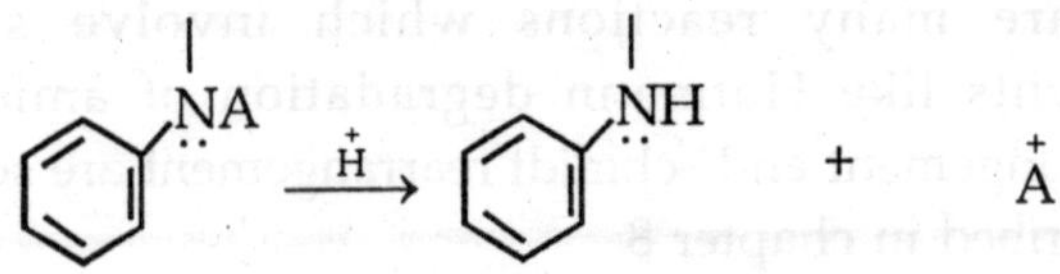

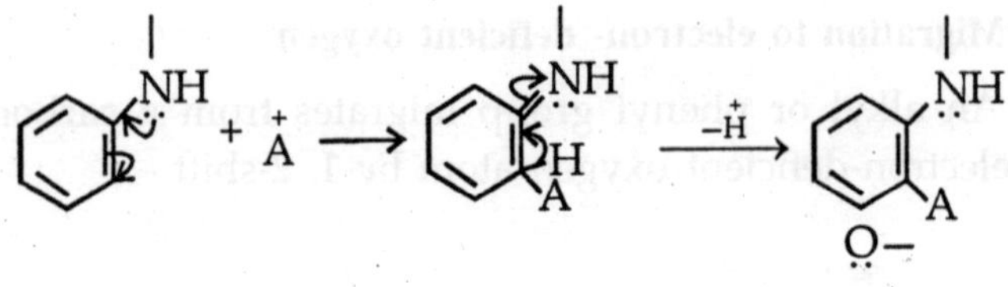

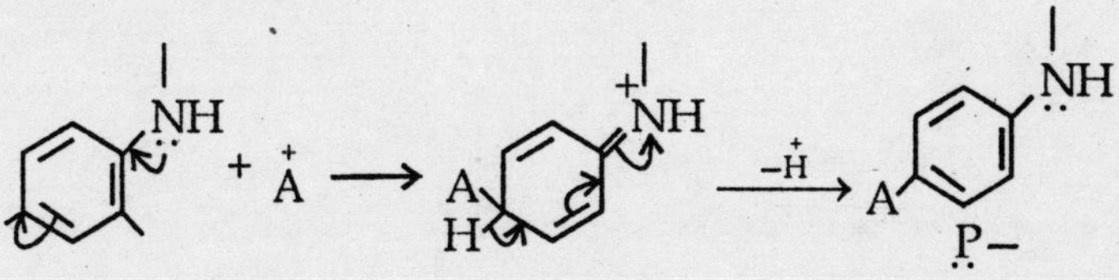

A few examples of such rearrangements are illustrated in chapter 8. viz. *rearrangement of diazoaminobenzene* to *p-aminoozobenzene, Fischer-Hepp rearrangement, Hofmann-Martius rearrangement* and *rearrangement of phenyl hydroxylamine* to *p- aminophenol.*

Chapter 8

REACTION MECHANISMS-II

Common principles involved in the mechanism of various classes of organic reactions have been already discuessed and described in chapter 7. In the present chapter, mechanisms of certain useful organic reactions are given.

1. Wurtz reaction: When as alkyl halide (R–X) is treated with sodium in dry ether, an alkane (R–R) is formed.

$$R{-}X + 2Na + R{-}X \xrightarrow{\text{ether}} R{-}R + 2NaX$$

If the alkyl halide is of two or more carbon atoms, an olefin and another alkane (R–H) are also produced as side products:

$$CH_3CH_2Br \xrightarrow{\text{Na, ether}} CH_3CH_2CH_2CH_3 + CH_2{=}CH_2 + CH_3CH_3$$

Mechanism: Mechanism of Wurtz reaction is not fully understood. Two probable mechanisms are given below:

(i) Mechanism involving organo-sodium compound formation.

Step I $\quad CH_2CH_3Br + 2Na \longrightarrow CH_3\overset{-}{C}H_2\overset{+}{N}a + NaBr$

Step II $\quad CH_3\overset{-}{C}H_2\overset{+}{N}a + CH_3CH_2Br \longrightarrow CH_3CH_2CH_2CH_3 + NaBr$

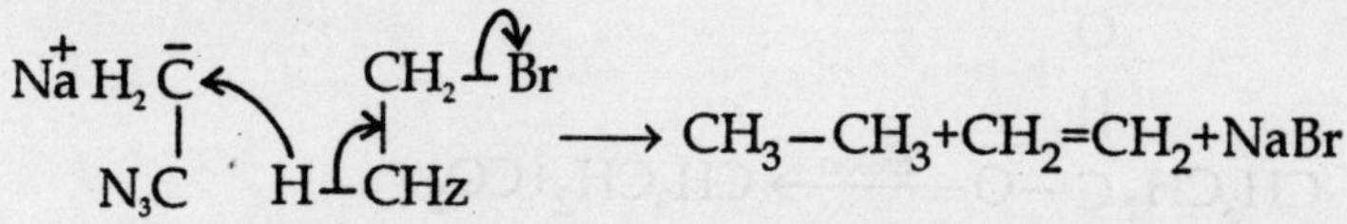

(ii) Mechanism involving free radical formation.

Step I $CH_2CH_3-Br+Na \longrightarrow CH_3CH_2+NaBr$

Step II $CH_3CH_2+CH_2CH_3 \xrightarrow{\text{coupling}} CH_3CH_2CH_2CH_3$

$$CH_3CH_2+CH_2CH_3 \xrightarrow[\text{tionation}]{\text{dispropor}} CH_2{=}CH_2CH_3CH_3$$

2. Kolbe Reaction: Electrolysis of concentrated solution of sodium or potassium salt of a fatty acid (RCOONa or RCOOK), produces alkane (R–R) at the anode.

$$2\ RCOONa+2H_2O \xrightarrow[\text{electrolysis}]{\text{Pt anode}} R-R+2CO_2+H_2+2NaOH$$

By-products such as olefins, lower alkanes etc. are also formed.

Mechanism:

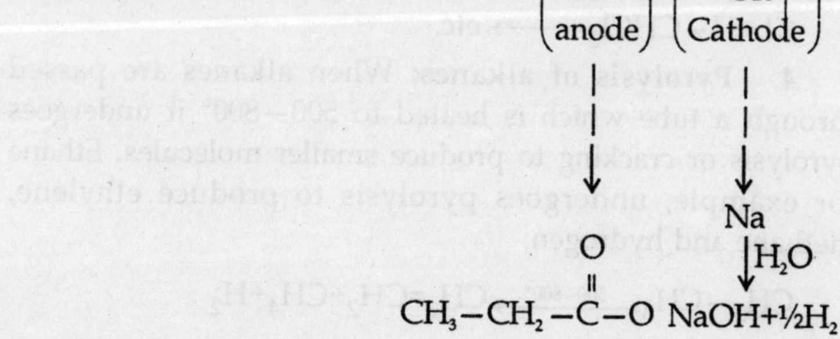

$$CH_3CH_2\overset{\overset{\displaystyle O}{\|}}{C}-O \xrightarrow{\text{fission}} CH_3CH_2 + CO_2$$

$$CH_3CH_2. + .CH_2CH_3 \xrightarrow{\text{coupling}} CH_3-CH_2-CH_2-CH_3$$

$$2CH_3-CH_2. \xrightarrow[\text{tlonation}]{\text{dispropor-}} CH_2{=}CH_2 + CH_3-CH_3$$

$$CH_3\text{-}CH_2 + CH_3\text{-}CH_2\text{-}\overset{\overset{\displaystyle O}{\|}}{C}-O \xrightarrow{\text{coupling}} CH_3\text{-}CH_2\text{-}\overset{\overset{\displaystyle O}{\|}}{C}\text{-}O\text{-}CH_2\text{-}CH_3$$

3. Halogenation of alkanes: At room temperature treating alkanes with chlorine or bromine in sunlight undergo *photohalogenation* substituting hydrogen atoms by halogen atoms.

$$CH_4 \xrightarrow[hv]{Cl_2} CH_3Cl \xrightarrow[hv]{Cl_2} CH_2Cl_2 \xrightarrow[hv]{Cl_2} CHCl_3 \xrightarrow[hv]{Cl_2} CCl_4$$

Mechanism: The reaction proceeds by a free radical chain mechanism. Once the reaction is initiated by homolytic fission of Cl–Cl bond by absorption of light, the chain goes on.

$$Cl_2 \xrightarrow{hv} Cl. + Cl.$$
$$CH_4 + Cl. \longrightarrow CH_3. + HCI$$
$$.CH_3 + Cl_2 \longrightarrow CH_3Cl + .Cl$$
$$Cl. + H-CH_2Cl \longrightarrow .CH_2Cl + HCl.$$
$$.CH_2Cl + Cl_2 \longrightarrow CH_2Cl_2 + .Cl$$
$$Cl. + H-CHCl_2 \longrightarrow \text{etc.}$$

4. Pyrolysis of alkanes: When alkanes are passed through a tube which is heated to 500–800° it undergoes pyrolysis or cracking to produce smaller molecules. Ethane for example, undergoes pyrolysis to produce ethylene, methane and hydrogen:

$$CH_3-CH_3 \xrightarrow{500-800^\circ} CH_2{=}CH_2 + CH_4 + H_2$$

Mechanism: Pyrolysis proceeds through a free radical mechanism involving following steps—

(i) *initiation*

$$H_3C_2CH_3 \xrightarrow{\text{heat}} H_3C.+.CH_3$$

(ii) *propagation*

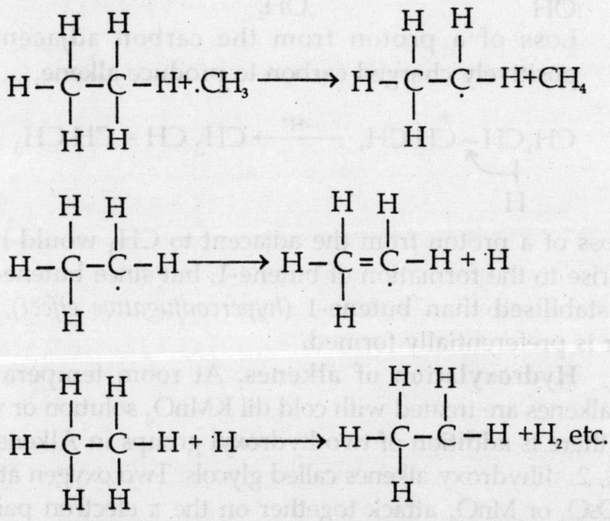

5. **Dehydrohalogenation of alkyl halides:** The alkyl halides produce alkenes on boiling with ethanolic KOH. This is an elimination reaction. See pages 83—55 for details.

6. **Dehydration of alcohols:** In presence of electron-acceptors* or proton-donors** , alcohols undergo dehydration by the process of elimination to produce corresponding olefins. The double bond is created between the carbon atom of C—OH and that adjacent carbon atom which is poorer in hydrogen (*Saytzeff's Rule*).

$CH_3-CH_2-CH-CH_3$ (with OH on the CH) $CH_3-CH{=}CH-CH_3$

butanol – 2   butene-2

---

* Electron-acceptors: alumina, P_2O_5, $ZnCl_2$ etc.

** Proton-donors: H_2SO_4, H_2PO_4 etc.

Mechanism: Dehydration of alcohol is an elimination reaction. Its mechanism involves following two steps:

I. Protonation of alcohol with subsequent loss of H_2O to produce a carbonium ion.

$$CH_3CH_2\underset{:OH}{\underset{|}{CH}}CH_3 \xrightarrow{H^+} CH_3CH_2\underset{_+OH_2}{\underset{|}{CH}}\ CH_3 \xrightarrow{-H_2O} CH_3CH_2\overset{+}{C}HCH_3$$

II. Loss of a proton from the carbon adjacent to positively charged carbon to produce alkene.

$$CH_3\underset{H}{\underset{|}{CH}}-\overset{+}{C}H\,CH_3 \xrightarrow{-H^+} CH_3\,CH=CH\,CH_3$$

Loss of a proton from the adjacent to CH_3 would have given rise to the formation of butene-1, but since butene-2 is more stabilised than butene-1 (*hyperconjugative effect*), the former is preferentially formed.

7. Hydroxylation of alkenes. At room temperature when alkenes are treated with cold dil $KMnO_4$ solution or with OsO_4, there is addition of two hydroxyl groups in Alkenes to form 1, 2, dihydroxy alkenes called glycols. Two oxygen atoms from OsO_4 or MnO_4 attack together on the π electron pair of the double bond of of alkene forming two bonds with two carbon atoms on the same side bringing about *cis* addition. The central metal atom Os or Mn acts as an electro-philic centre and is reduced by accepting extra electrons.

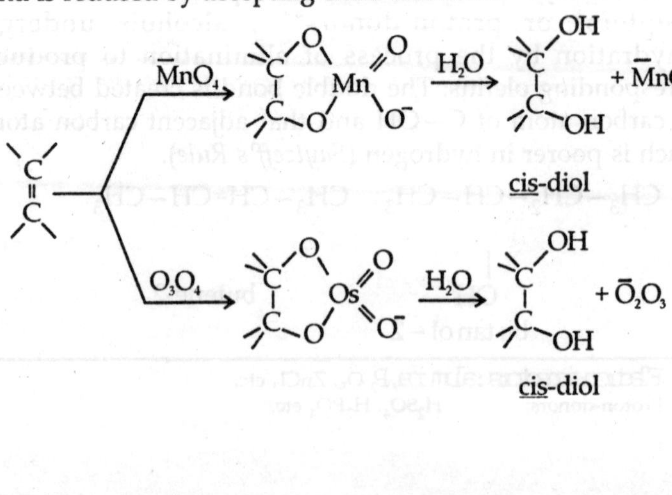

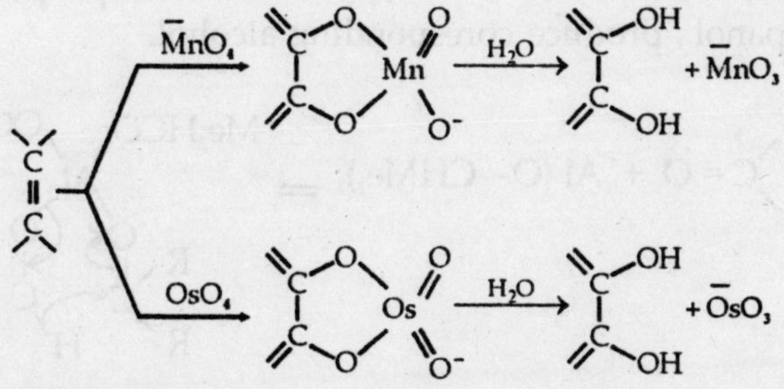

Cis-diol

8. Ozonisation of alkenes: Ozone interacts with both σ as well as π electron pairs of double bond of alkenes to form ozonides. Because of their explosive nature, the ozonides are not isolated. Decomposition of ozonides produces carbonyl compounds.

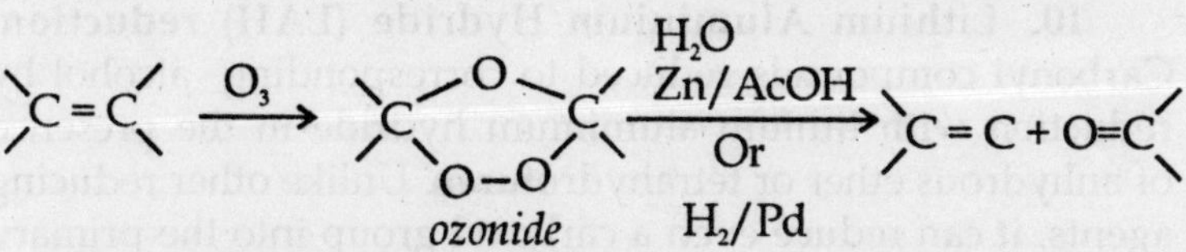

Mechanism: Ozone behaves as an electrophile, as is evident from its resonating structures:

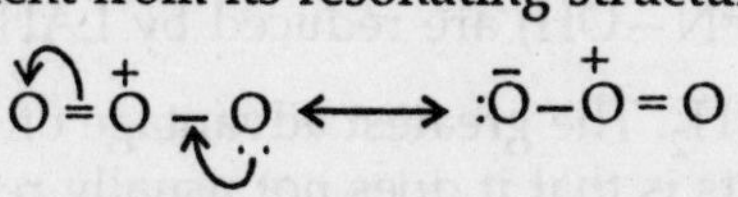

Formation of ozonide mechanism involves following steps starting with *cis* addition of O_3 to alkene:

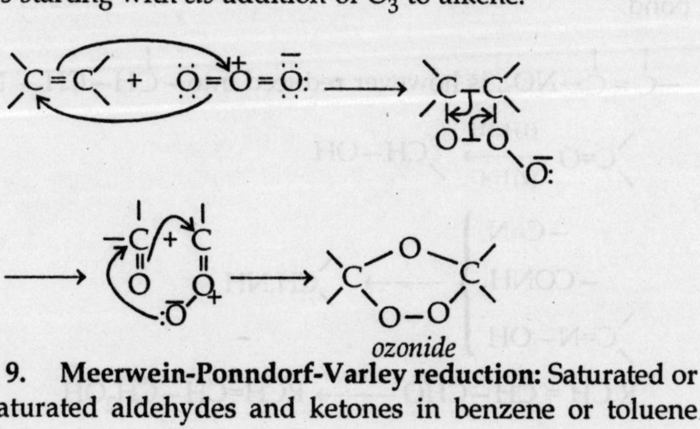

9. Meerwein-Ponndorf-Varley reduction: Saturated or unsaturated aldehydes and ketones in benzene or toluene solution, when treated with aluminium isopropoxide in isopropanol., produce coresponding alcohol.

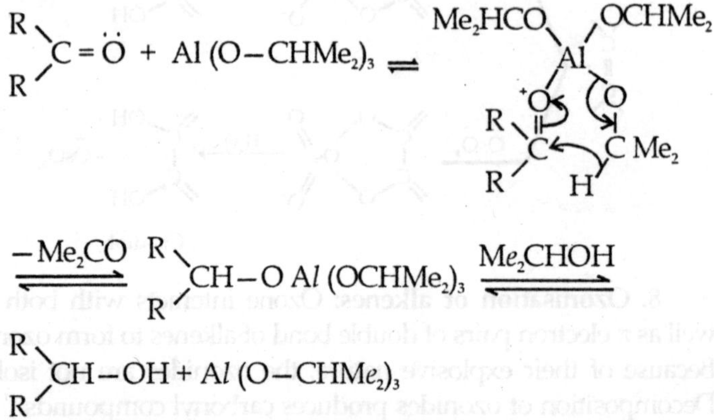

10. Lithium Aluminium Hydride (LAH) reduction: Carbonyl compounds reduced to corresponding alcohol by reduction with lithium aluminium hydride in the presence of anhydrous ether or tetrahydrofuran. Unlike other reducing agents, it can reduce even a carboxyl group into the primary alcohol. Other functional groups like nitriles (–CN), amides ($-CONH_2$), oximes ($>C=N-OH$) are reduced by LAH into primary amines $-CH_2NH_2$. The greatest advantage of LAH over other reducing agents is that it does not usually reduce an olefinic bond. Nitro group in conjugation with an olefinic bond.

$-\overset{|}{C}=\overset{|}{C}-NO_2$, is however reduced into $-\overset{|}{C}H-CH_2-NH_2$.

$$>C=O \xrightarrow[(2)\ H_2O]{(1)\ LAH} >CH-OH$$

$$\left.\begin{array}{r} -C\equiv N \\ -CONH_2 \\ >C=N-OH \end{array}\right\} \longrightarrow\ >CH_2NH_2$$

$$RCH=CH-CHO \longrightarrow RCH=CH-CH_2OH$$

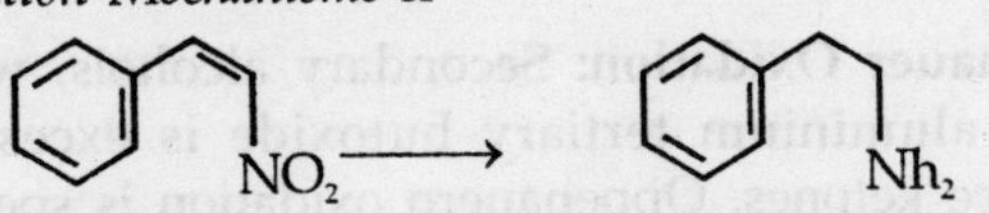

Mechanism:

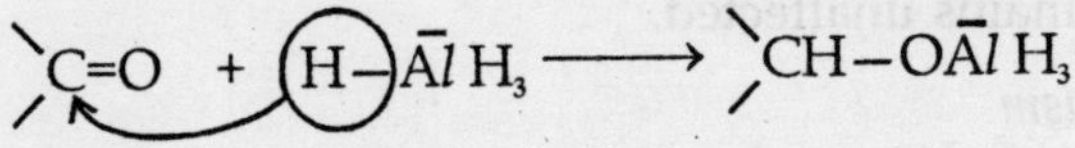

$$\xrightarrow{>C=O} \left(>CH-O\right)_2\bar{Al}H_2 \xrightarrow{>C=O} \left(>CH-O\right)_3 Al\,H \xrightarrow{>C=O}$$

$$\left(>CH-O\right)_4\bar{Al} \xrightarrow{H_2O} 4>CH-OH + Al(OH)_3$$

11. Esterification: When carboxylic acid is heated with alcohols in the presence of few drops of acid(5-10% conc. H_2SO_4), forms corresponding ester.

$$R-\overset{O}{\overset{\|}{C}}-\boxed{OH+H}-OR \underset{}{\overset{H^+}{\rightleftharpoons}} R-\overset{O}{\overset{\|}{C}}-OR+H_2O$$

Mechanism:

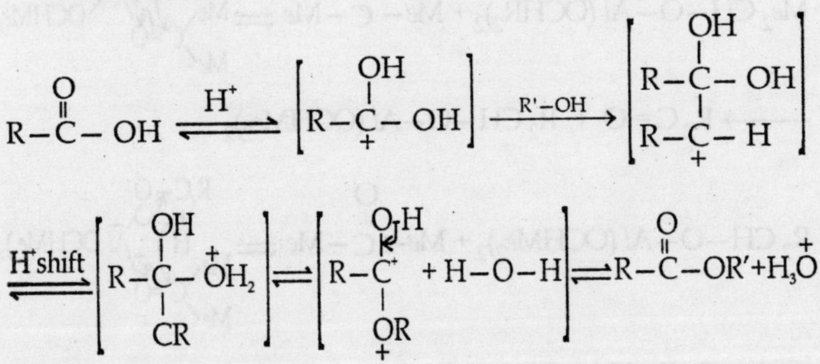

An alcohol when treated with halogen acid in presence of a Lewis acid forms corresponding ester (alkyl halide) by a different mechanism.

$$R-\boxed{OH+H}-X \longrightarrow R-X+H_2O$$

Mechanism:

$$R-\ddot{O}H+HX \longrightarrow R-\overset{+}{O}H_2 + :\bar{X}$$

$$R-\underset{+}{\overset{H}{\overset{|}{O}}}-H \longrightarrow R^+ + H-\ddot{O}-H$$

$$\bar{R}-\bar{X} \longrightarrow R-X$$

12. Oppenauer Oxidation: Secondary alcohols, when refluxed with aluminium tertiary butoxide is excess of acetone, produce ketones. Oppenauero oxidation is specific for secondary alcoholic group; the unsaturation in the molecule remains unaffected.

Mechanism

$$3R_2CHOH + Al(OCMe_3)_3 \rightleftharpoons (R_2CH{-}O)_3Al + 3Me_3C{-}OH$$

$$(R_2CH{-}O)_3Al + Me{-}\overset{O}{\overset{\|}{C}}{-}Me \rightleftharpoons$$

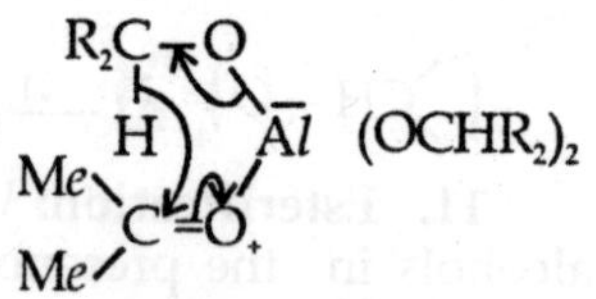

$$\longrightarrow R_2C{=}O + Me_2CH{-}O{-}Al(OCHR_2)_2$$

$$Me_2CH{-}O{-}Al(OCHR_2)_2 + Me{-}\overset{O}{\overset{\|}{C}}{-}Me \rightleftharpoons$$

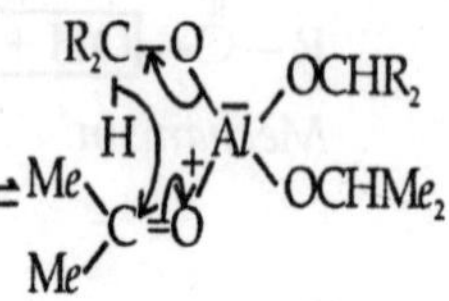

$$\longrightarrow R_2C{=}O + R_2CH{-}O{-}Al(OCHMe_2)_2$$

$$R_2CH{-}O{-}Al(OCHMe_2)_2 + Me{-}\overset{O}{\overset{\|}{C}}{-}Me \rightleftharpoons$$

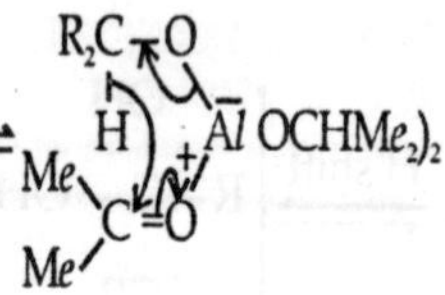

$$\longrightarrow R_2C{=}O + Al(OCHMe_2)_3$$

13. Pinacol-pinacolone rearrangement: 2, 3-dimethyl-butane 2, 3-diol (*pinacol*) when distilled with H_2SO_4produces methyl *tert*-butyl ketone (*pinacolone*).

$$\underset{\text{Pinacol}}{(CH_3)_2{-}\underset{OH}{\underset{|}{C}}{-}\underset{OH}{\underset{|}{C}}{-}(CH_3)_2} \xrightarrow{H^+} \underset{\text{Pinacolone}}{(CH_3)_3C{-}\underset{O}{\underset{\|}{C}}{-}CH_3}$$

Mechanism: One out of two OH groups of pinacol is protonated with subesequent loss of a water molecule to form a carbonium ion. This carbonium ion then rearranges by 1, 2-alkyl shift into a more stable carbonium ion. Then proton is lost from the other OH group results in the formation of a ketone.

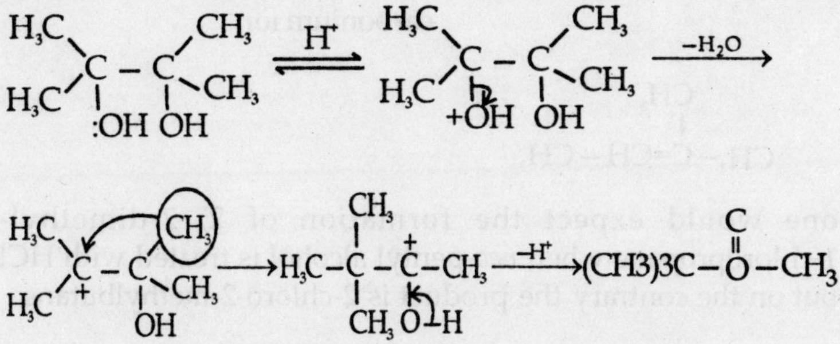

14. Wagner Meerwein rearrangement: Dehydration of neopentyl alcohol in acidic condition produces 2-methylbutene-2.

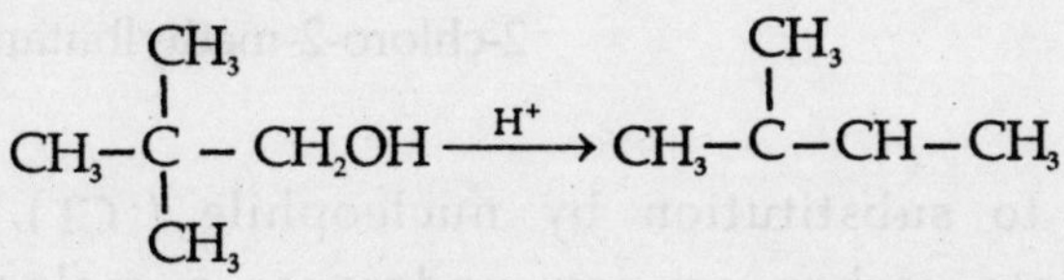

Mechanism: The protonated alcohol results in neopentyl carbonium on that rearranges by 1, 2-methyl shift to a more stable 2-methyl-2-butyl carbonium ion with subsequent loss of proton from the carbon adjacent to $-\overset{|+}{C}-$.

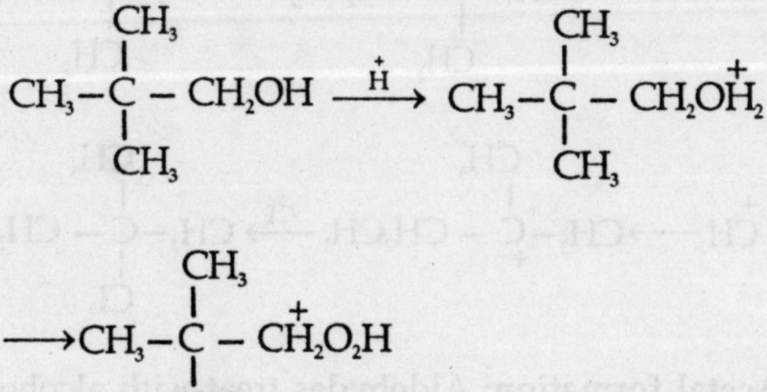

Neo-pentyl carbonium ion

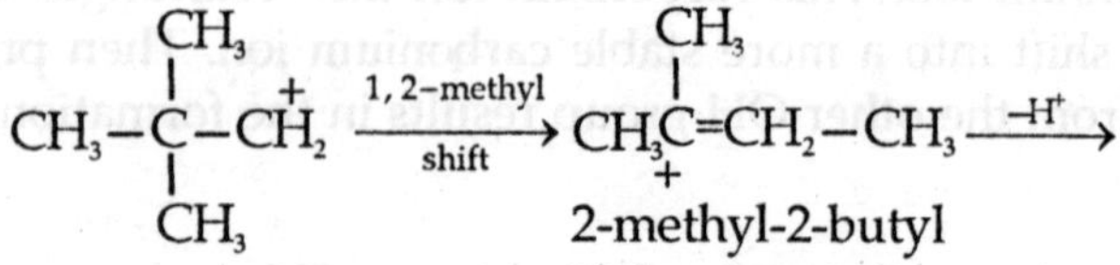

2-methyl-2-butyl carbonium ion

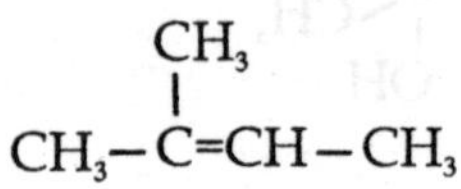

one would expect the formation of 2, 2-dimethyl-1-chloropropane when *neo*-pentyl alcohol is treated with HCl but on the contrary the product is 2-chloro-2-methylbutane.

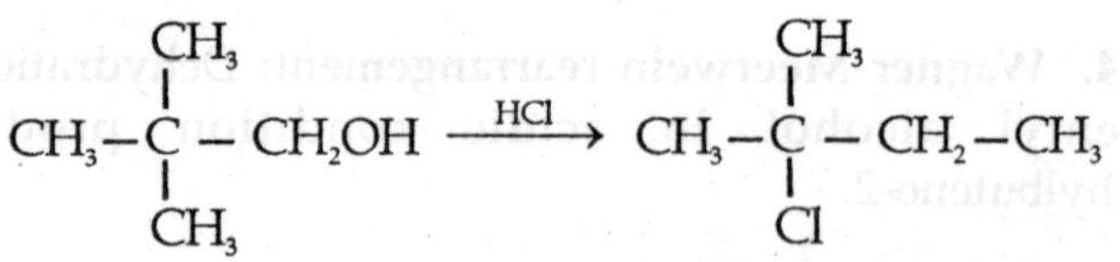

2-chloro-2-methylbutane

Prior to substitution by nucleophile ($:\bar{Cl}$), the intermediate carbonium ion undergoes a molecular rerrangement by 1, 2-methyl-shift into a more stable carbonium ion:

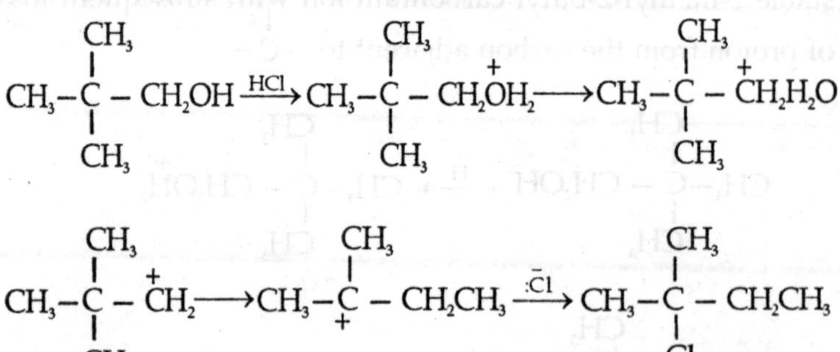

15. Acetal formation: Aldehydes treat with alcohols in presence of dry HCL gas to form acetal.

Mechanism:

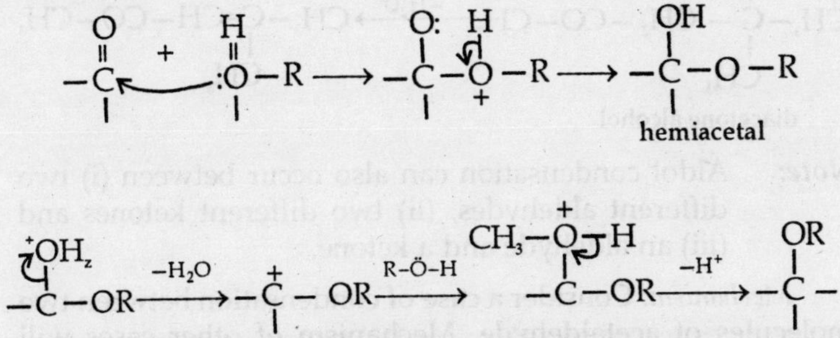

16. Aldol condensation: Aldehydes or ketones, having α-hydrogen undergo self-condensation in presence of a base catalyst to produce α-hydroxy- aldelhydes or ketones, called *aldols*. On heating these aldols undergo dehydration to produce α, β-unsaturated aldehydes or ketones A few specific examples are cited below:

(i) Two molecules of acetaldehyde undergo condensation to produce β-hydroxybutyraldehyde (aldol) in presence of dil. NaOH. Aldol loses a molecule of water on heating, to give crotonaldehyde—

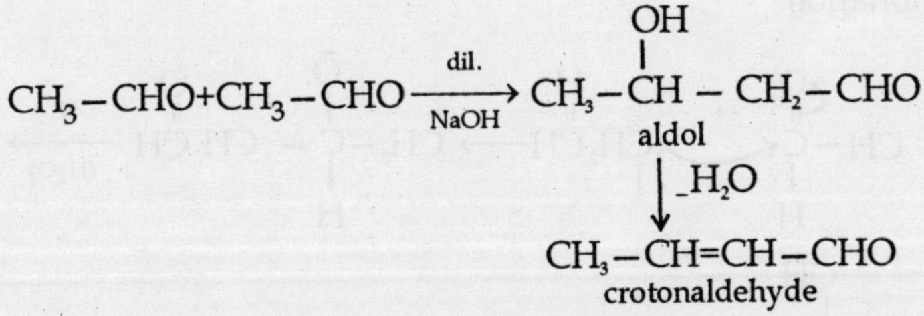

(ii) Two molecules of acetone undergo condensation to produce diacetone alcohol in presence of aqueous barium hydroxide. Diacetone alcohol loses a molecule of water in presence of a little of iodine to give mesityl oxide.

$$CH_3-CO-CH_3+CH_3-CO-CH_3 \xrightarrow{Ba(OH)_2}$$

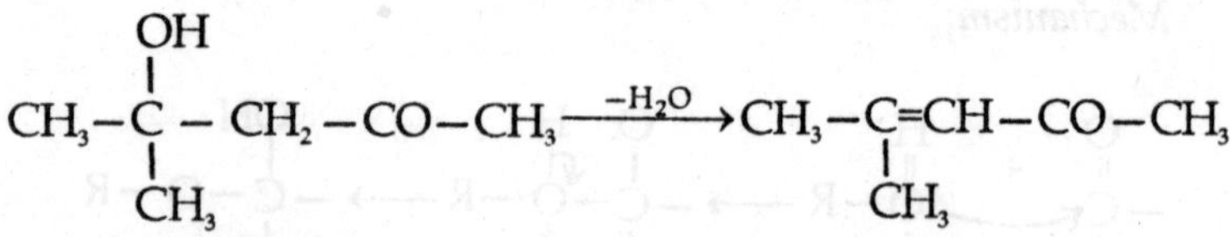

diacetone alcohol

Note: Aldol condensation can also occur between (i) two different aldehydes, (ii) two different ketones and (iii) an aldehyde and a ketone.

Mechanism: Consider a case of condensation between two molecules of acetaldehyde. Mechanism of other cases will follow a similar route.

Step I—The base abstracts a proton from α position of one of the acetaldehyde molecules:

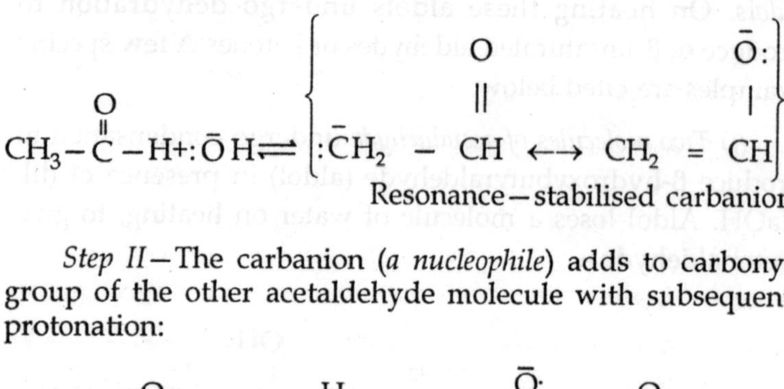

Resonance—stabilised carbanion

Step II—The carbanion (*a nucleophile*) adds to carbonyl group of the other acetaldehyde molecule with subsequent protonation:

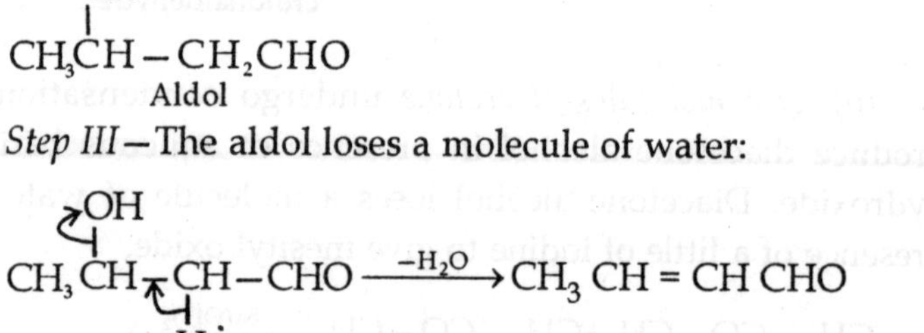

Aldol

Step III—The aldol loses a molecule of water:

crotonaldchyde

17. Claisen-Schmidt reaction (*Claisen reaction*): Aliphatic aldehydes and ketones undergo aldol-type condensation with aromatic aldehydes or ketones in presence of dil. NaOH to form aldol-type products, subsequent dehydration of β-unsaturated side chain aromatic aldehydes or ketones.

For example, benzaldehyde and acetaldehyde condense in presence of dil. NaOH to form a β–hydroxy aldehyde that on heating gives cinnamaldehyde:

$$C_6H_5-\overset{O}{\overset{\|}{C}}H + H-CH_2-CHO \xrightarrow[\text{NaOH}]{\text{dil.}} C_6H_5-\overset{OH}{\overset{|}{C}H}-CH_2CHO \xrightarrow[\text{aldol}]{(-H_2O)} \underset{\text{Cinnamaldehyde}}{C_6H_5-CH{=}CH-CHO}$$

Mechanism:

$$CH_3CHO \xrightarrow{:\bar{O}H} :\bar{C}H_2CHO + H_2O$$

$$C_6H_5-\overset{O}{\overset{\|}{C}}H + :CH_2CHO \longrightarrow C_6H_5\overset{\bar{O}:}{\overset{|}{C}}HCH_2CHO \xrightarrow[(H_2O)]{H^+}$$

$$C_6H_5\overset{OH}{\overset{|}{C}H}\underset{H}{\underset{|}{C}H}CHO \xrightarrow{-H_2O} C_6H_5CH=CH\,CHO$$

18. Knoevenagel reaction: In presence of an organic base such as pyridine or diethylamine, aldehydes react with compounds having a reactive methylene group in the molecule to produce aldol-type product, which on heating undergoes dehydration to produce β, α–unsaturated compound.

$$R-\overset{O}{\overset{\|}{C}}-H + CH_2(COOEt)_2 \xrightarrow{\text{Pyridine}} R-\overset{OH}{\overset{|}{C}H}-CH(COOEt)_2$$

$$\xrightarrow[-H_2O]{heat} RCH{=}C(COOEt)_2$$

Mechanism:

$$CH_2(COOEt)_2 \xrightarrow{C_5H_5N} :\bar{C}H(COOEt)_2 + C_5H_5\overset{+}{N}H$$

$$R\overset{O}{\overset{\|}{C}}H + :\bar{C}H(COOEt)_2 \longrightarrow R\overset{\bar{O}:}{\overset{|}{C}}H\,CH(COOEt)_2$$

$$\xrightarrow[(C_5H_5\overset{+}{N}H)]{H+} R\overset{OH}{\overset{|}{C}H}-\underset{H}{\underset{|}{C}}(COOEt)_2 \longrightarrow RCH{=}C(COOEt)_2$$

19. Cannizzaro Reaction: Aldehydes, having no α hydrogen, In the presence of 60% KOH, aldehydes disproportionate which result in the formation of an acid anion and a primary alcohol. In this process of disproportionation, one molecule of the aldenyde is reduced at the cost of Oxidation of the other.

For *example*:

(i) Two molecules of formaldehyde undergo Cannizzaro reaction to form formate anion and methyl alcohol.

$$H-\overset{O}{\overset{\|}{C}}-H + H-\overset{O}{\overset{\|}{C}}-H \xrightarrow[KOH]{60\%} HCO\bar{O}\,\bar{K} + CH_3OH$$

(ii) Formaldehyde and benzaldehyde undergo Cannizzaro reaction to form formate anion and benzyl alcohol. Here benzaldehyde is reduced at the cost of formaldehyde (Crossed Cannizzaro reaction).*

---

* The nucleophilic attack on the carbon atom of carbonyl group is easier in formaldehyde than in benaldehyde. This is due to -(ii)the steric hindrance caused by the bulky phenyl group in benzaldehyde and (ii)the diminished positive charge on the carbon atom of the carbonyl group of benzaldehyde because of the extension of resonance of the ring to the carbonyl group.

$$C_6H_5-\overset{O}{\overset{\|}{C}}-H + H-\overset{O}{\overset{\|}{C}}-H \xrightarrow[\text{KOH}]{60\%} C_6H_5CH_2OH + HCO\bar{O}\overset{\pm}{K}$$

Mechanism:

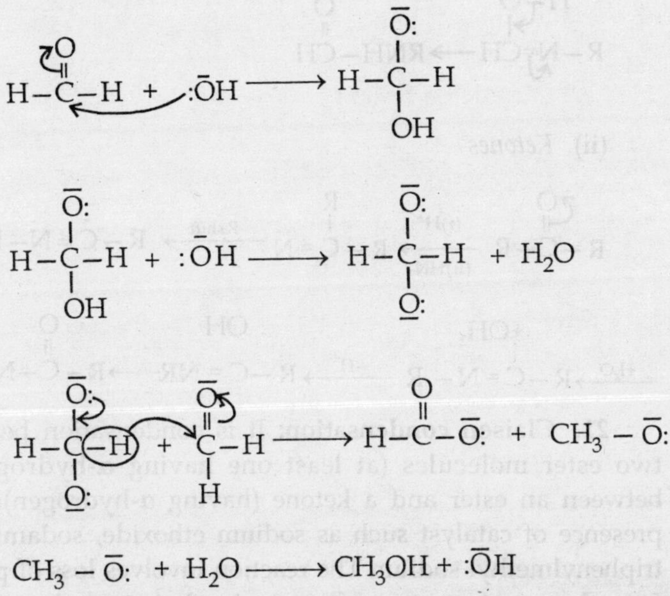

20. Schmidt reaction: An aldehyde (RCHO) reacts with hydrazoic acid in presence of conc. H_2SO_4 to produce an alkyl cyanide (RCN) and formyl derivative of primary amine (RNHCHO). On the other hand, ketones produce acyl derivatives of *primary* amine (R—NHCOR).

(i) *Aldehydes*

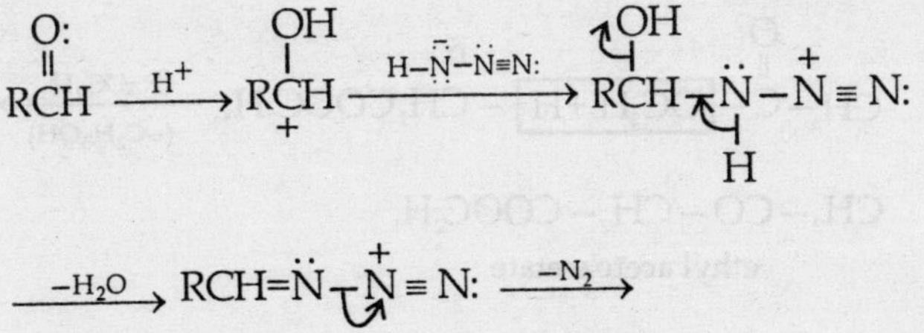

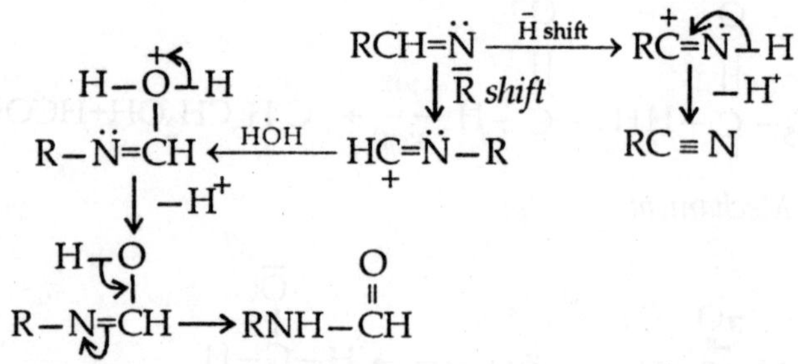

(ii) *Ketones*

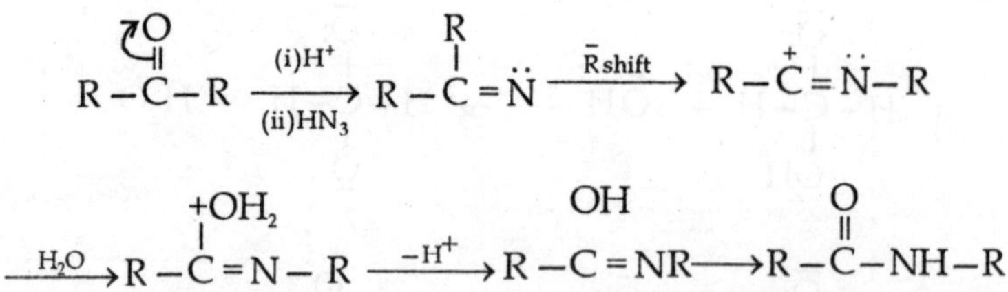

21. Claisen condensation: It is condensation between two ester molecules (at least one having α-hydrogen)or between an ester and a ketone (having α-hydrogen)in the presence of catalyst such as sodium ethoxide, sodamide or triphenylmethyl sodium The reaction involves loss of proton from the α-carbon atom of one molecule (*ester or ketone*) and alkoxy anion from the other (*ester*) with subsequent union of the residual units to produce a β-keto ester or a 1, 3-diketone as the case may be.

Examples:

(i) Condensation between two identical ester molecules—

$$CH_3-\overset{O}{\overset{\|}{C}}-\boxed{OC_2H_5+H}-\overset{\alpha}{C}H_2COOC_2H_5 \xrightarrow[(-C_2H_5OH)]{NaOC_2H_5}$$

$$\underset{\text{ethyl acetoacetate}}{CH_3-CO-CH_2-COOC_2H_5}$$

(ii) Condensation between two different ester molecules–

$$\underset{\text{}}{H-\overset{\overset{\displaystyle O}{\|}}{C}-\boxed{OC_2H_5+H}\overset{\alpha}{-}CH_2COOC_2H_5} \xrightarrow[(-C_2H_5OH)]{NaOC_2H_5}$$

$$H-\overset{\overset{\displaystyle O}{\|}}{C}-CH_2-COOC_2H_5$$

ethyl formoacetate

(iii) Condensation between an ester and a ketone molecule–

$$CH_3-\overset{\overset{\displaystyle O}{\|}}{C}-\boxed{OC_2H_5+H}\overset{\alpha}{-}CH_2-\overset{\overset{\displaystyle O}{\|}}{C}-CH_3 \xrightarrow[(-C_2H_5OH)]{NaOC_2H_5}$$

$$CH_3-\overset{\overset{\displaystyle O}{\|}}{C}-CH_2-C-CH_3$$

Pentanedione-2, 4

Most probable mechanism for Claisen condensation, taking the case of ethyl acetate, involves following steps–

Step I: Removing of a proton from the α carbon of ester molecule by the ethoxide ion to form a carbanion.

$$CH_3-\overset{\overset{\displaystyle O}{\|}}{C}-OC_2H_5 + :\bar{O}C_2H_5 \longrightarrow \bar{C}H_2-\overset{\overset{\displaystyle O}{\|}}{C}-OC_2H_5 + C_2H_5OH$$

carbanion

Step II: The nucleophilic carbanion attacks the carbonyl carbon atom of the other ester molecule displacing ethoxide ion.

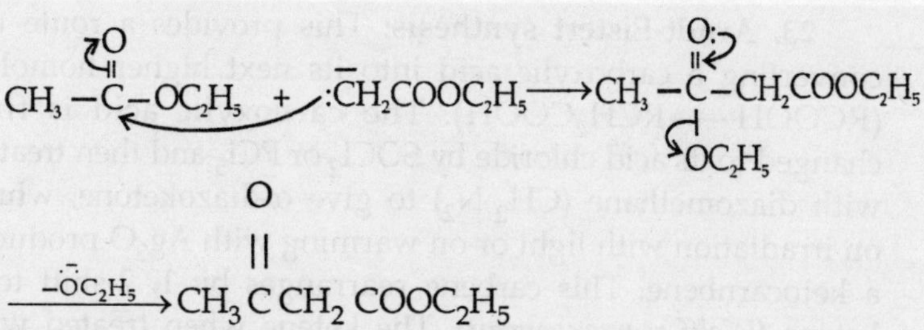

22. Baeyer Villiger oxidation: In this reaction, using a peracid an aliphatic ketone is converted into an ester.

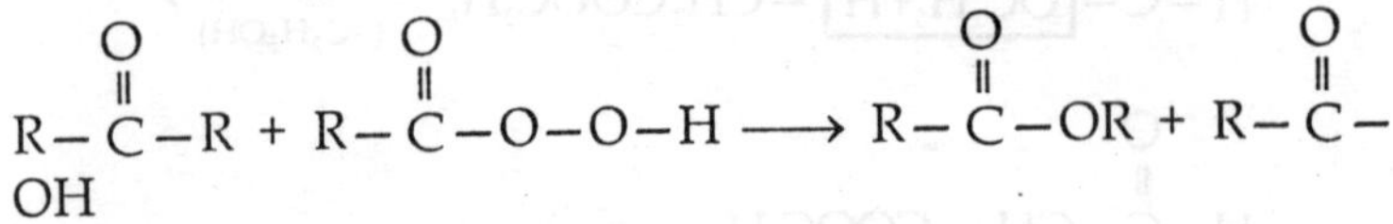

Mechanism:

In the first step there is addition of peracid to the carbonyl group of ketone In second step, the heterolytic breaking of O–O linkage of the adduct is followed by the *rearrangement* of an alkyl group to the electron- deficient oxygen atom.

I Step:

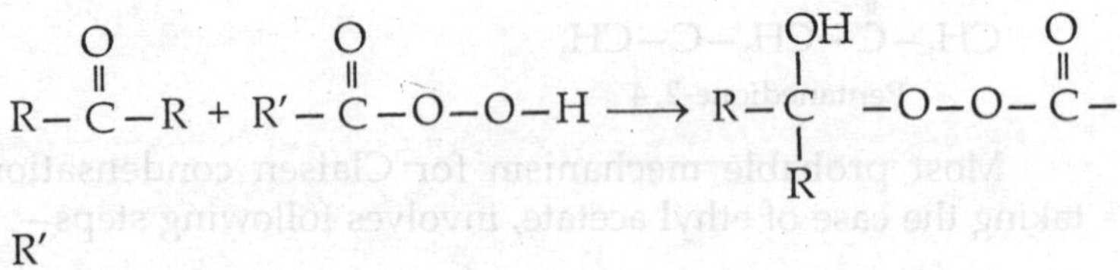

II Step:

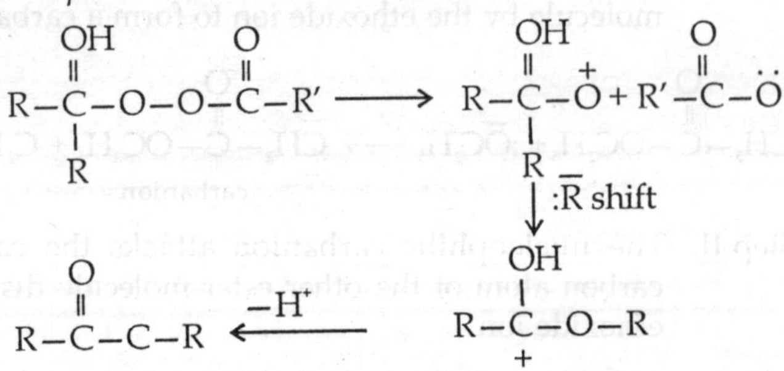

23. Arndt-Eistert synthesis: This provides a route for converting a carboxylic acid into its next higher homolog ($RCOOH \longrightarrow RCH_2COOH$). The carboxylic acid is first changed to its acid chloride by $SOCl_2$ or PCl_5 and then treated with diazomethane (CH_2 N_2) to give α-diazoketone, which on irradiation with light or on warming with Ag_2O produces a ketocarnbene. This carbene rearranges by 1, 2-shift to a ketene (*Wolff rearrangement*). The ketene when treated with water yields carboxylic acid.

$$R-\overset{\overset{O}{\|}}{C}-OH \xrightarrow[\text{or } PCl_3]{SOCl_2} R-\overset{\overset{O}{\|}}{C}-Cl$$

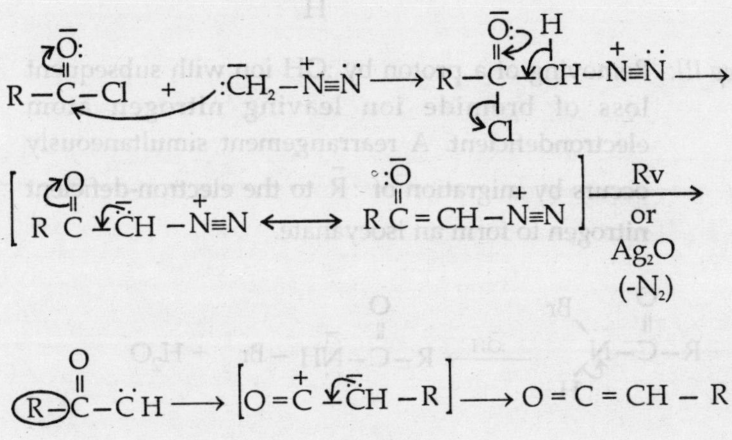

$$\xrightarrow{HOH} RCH_2COOH$$

24. Hofmann degradation of amides: Amides are converted to primary amines by means of bromine and NaOH.

$$RCONH_2 + Br_2 + 4NaOH \longrightarrow RNH_2 + 2NaBr + Na_2CO_3 + 2H_2O$$

or (in place of $RCONH_2$): Ar $CONH_2$; or (in place of RNH_2): Ar NH_2

Mechanism;

Step I: Formation of : $\bar{O}$ Br ion.

$$2NaOH + Br_2 \longrightarrow Na\,OBr + NaBr + H_2O$$

$$NaOBr \longrightarrow \overset{+}{N}a + :\bar{O}Br$$

Step II: Bromination of the amide by : $\bar{O}$ Br ion to form N-bromamide.

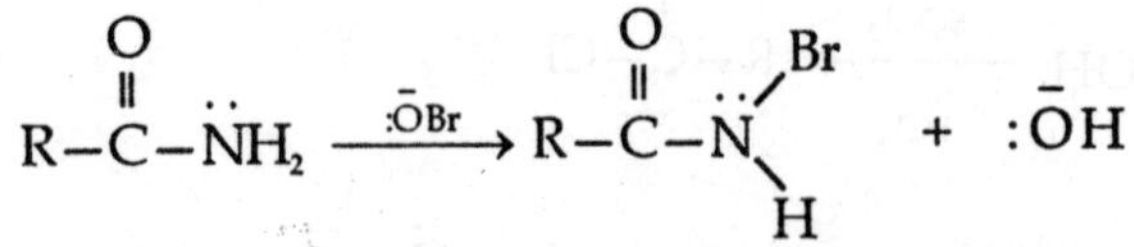

Step III: Removing of a proton by :OH ion with subsequent loss of bromide ion leaving nitrogen atom electrondeficient. A rearrangement simultaneously occurs by migration of $:\bar{R}$ to the electron-deficient nitrogen to form an isocyanate.

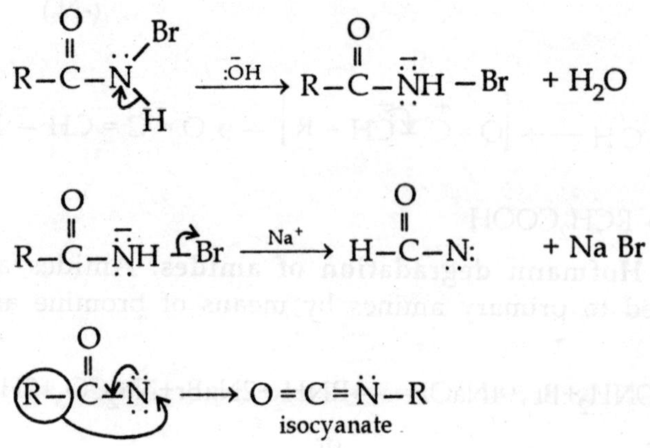

Step IV: Hydrolysis of isocyanate.

$$O{=}C{=}\ddot{N}-R \xrightarrow{H_2O} RNH-CO_2H \longrightarrow RNH_2 + CO_2$$

25. Curtius reaction: Using this reaction we can convert acyl halide to primary amine. The acyl halide is treated with sodium azide to get acyl azide. Acyl azide on heating evolves. nitrogen to give an isocyanate which on hydrolysis forms a primary amine.

$$RCOCl + NaN_3 \longrightarrow RCON_3 \xrightarrow{heat} \underset{\text{isocyanate}}{R-N{=}C{=}O} + N_2$$

$$R-N{=}C{=}O \xrightarrow{H_2O} RNH_2 + CO_2$$

Mechanisum:

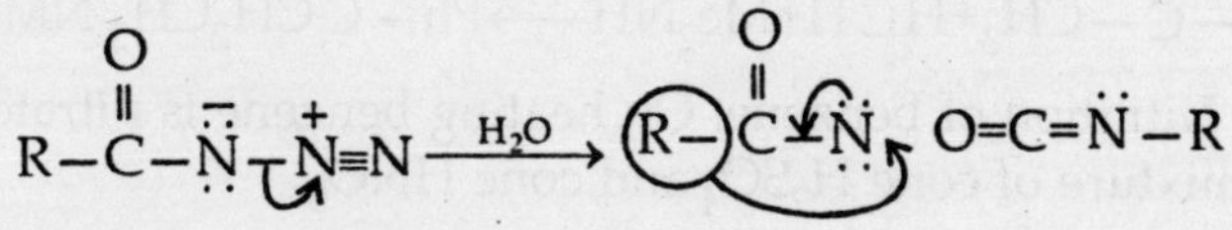

$\xrightarrow{HOH} RNH_2 + CO_2$

26. Mannich reaction: This reaction is used to convert $-\overset{O}{\overset{\|}{C}}-\overset{|}{C}H-$ linkage of an aldehyde or a ketone into $-\overset{O}{\overset{\|}{C}}-\underset{|}{\overset{|}{C}}-CH_2NH_2$ (*Manich base*). It involves condensation between formaldehyde, ammonia (or a primary or secondary amine) and an aldehyde or a ketone must have minimum of one α-hydrogen atom. The amine and formaldehyde first condense to form the *immonium* salt. The enolate of the carbonyl compound, formed by amine acting as a base, then attacks the immonium intermediate gives rise to *Mannich base*:

$$H-\overset{O}{\overset{\|}{C}}-H + \underset{\begin{pmatrix} R=H \\ R=alkyl \end{pmatrix}}{HNR_2} \rightleftharpoons \underset{\begin{pmatrix} \text{immonium} \\ \text{salt} \end{pmatrix}}{R_2\overset{+}{N}=CH_2} + :\bar{O}H$$

$$-\overset{O}{\overset{\|}{C}}-\overset{|}{C}H- \overset{\text{base}}{\rightleftharpoons} -\overset{\bar{O}:}{\overset{\|}{C}}=\underset{|}{C}-$$

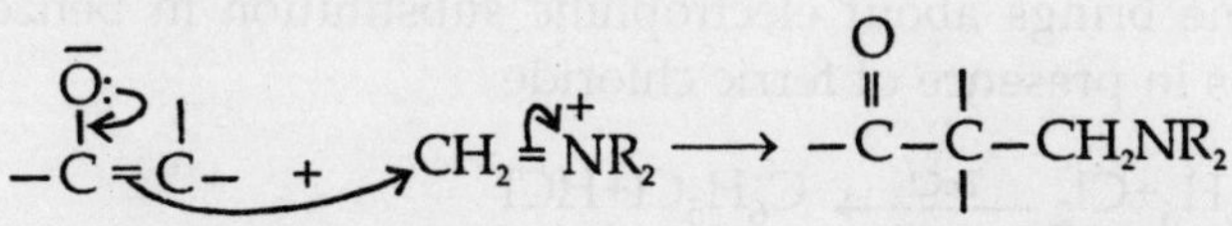

Mannich base

$$Ph-\overset{\overset{O}{\|}}{C}-CH_3 + H\overset{\overset{O}{\|}}{C}H + Me_2NH \longrightarrow Ph-\overset{\overset{O}{\|}}{C}CH_2CH_2\,NMe_2$$

27. Nitration of benzene: On heating benzene is nitrated with a mixture of cone H_2SO_4 and cone HNO_3

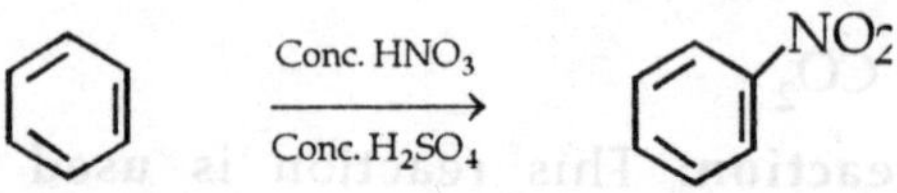

Mechanism:

Generation of the electrophile:

$$H_2SO_4 \rightleftharpoons H^+\,H\bar{S}O_4$$

$$H^+ + H\ddot{O}-NO_2 \longrightarrow H-\overset{\overset{H}{|}}{\underset{+}{O}}-NO_2 \longrightarrow \overset{+}{N}O_2 + H_2O$$

nitronium ion

Attack of the electrophile $\overset{+}{N}O_2$ on a π-electron pair of the benzene ring to form an intermediate carbonium ion which is stabilised is due to resonance.

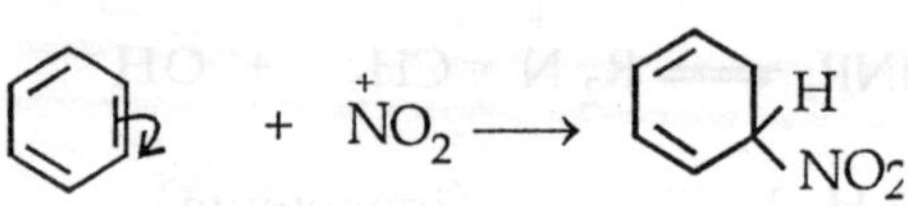

Abstraction of a proton from the carbonium ion:

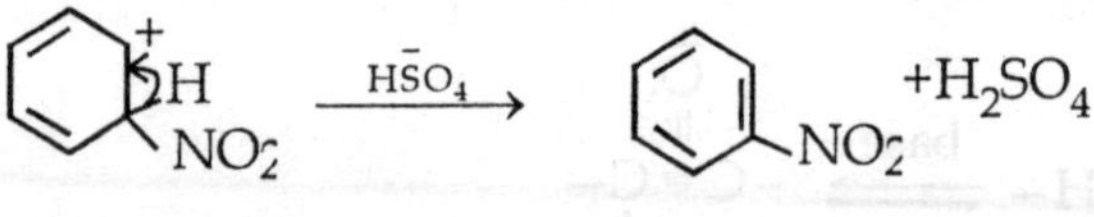

28. **Chlorination of benzene:** At room temperature Chlorine brings about electrophilic substitution in benzene nucleus in presence of ferric chloride

$$C_6H_6 + Cl_2 \xrightarrow{FeCl_2} C_6H_5Cl + HCl$$

Mechanism:

Benzene chlorination is an electrophilic substitution and advance by following mechanism:

$$Cl_2 + Fe\,Cl_3 \rightleftharpoons \overset{+}{Cl} + Fe\,Cl_4$$

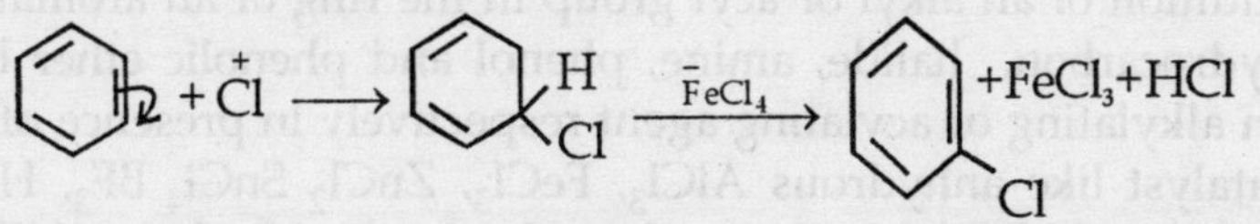

29. Sulfonation of benzene: When benzene is heated with concentrated sulfuric acid at 80°, undergoes substitution to produce benzenesulfonic acid.

$$C_6H_6 + H_2SO_4 \xrightleftharpoons{80^\circ} C_6H_5SO_3H + H_2O$$

Mechanism:

Sulfonation of benzene is an electrophilic substitution and advances by following mechanism:

$$2H_2\,SO_4 \rightleftharpoons SO_2 + H_3\overset{+}{O} + H\overset{-}{S}\,O_4$$

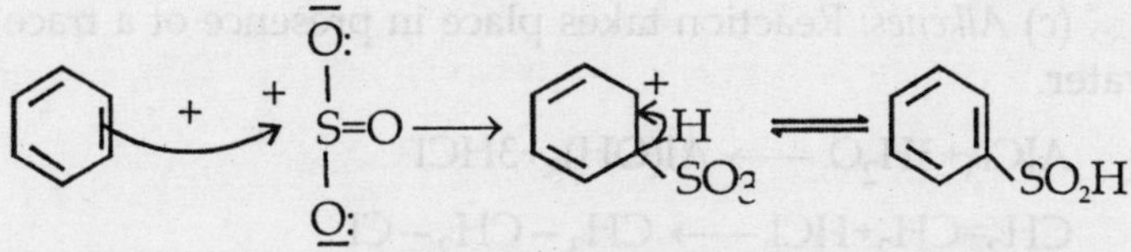

30. Wurtz-Fittig reaction: When an ethereal solution of an alkyl and aryl halide is warmed with sodium the alkyl group substitutes halogen is aryl halide to form alkyl derivative of the aromatic compound.

$$Ar-\boxed{X+2Na+X}-R \longrightarrow Ar-R+2\,NaX$$

By-products like Ar-Ar and R-R are also produced.

Mechanism:

$$Ar-X+Na. \longrightarrow Ar.+Na\text{-}X$$

$$R-X+Na \longrightarrow R. + Na\text{-}X$$

$$Ar. + R. \longrightarrow Ar-R$$

$$Ar. + Ar. \longrightarrow Ar-Ar$$

$$R. + R. \longrightarrow R-R$$

31. Friedel. Crafts reaction: This reaction involves addition of an alkyl or acyl group in the ring of an aromatic hydrocarbon, halide, amine, phenol and phenolic ether by an alkylating or acylating agent respectively in presence of a catalyst like anhydrous $AlCl_3$, $FeCl_3$, $ZnCl_2$ $SnCl_4$ BF_3, HF, H_2SO_4 or H_3PO_4. Mostly used catalysts are anhydrous $AlCl_3$ and BF_3.

(i) Alkylation—It is brought about by any of the following alkylating agents:

(a) *Alkyl halides:*

$$Ar-H+R-X \xrightarrow{AlCl_2} Ar-R+H-X$$

(b) *Alcohols:* In case of alcohols BF_3 catalyst is most effective.

$$Ar-H+R-OH \xrightarrow{BF_2} Ar-R+H_2O$$

(c) *Alkenes:* Reaction takes place in presence of a trace of water.

$$AlCl_3+3H_2O \longrightarrow Al(OH)_3+3HCl$$

$$CH_2{=}CH_2+HCl \longrightarrow CH_3-CH_2-Cl$$

$$Ar-H+CH_3CH_2Cl \xrightarrow{AlCl_3} Ar-CH_2CH_3+HCl$$

Mechanism:

$$R-Cl+AlCl_3 \longrightarrow \overset{\delta_+}{R}..Cl..\overset{\delta_-}{Al}Cl_2 \longrightarrow \overset{+}{R}+AlCl_4$$

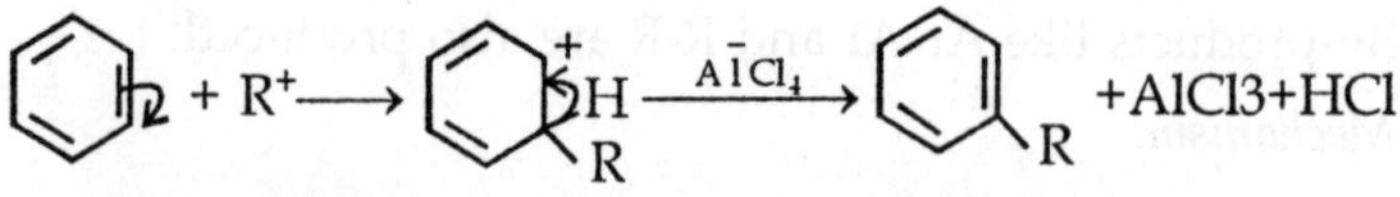

(ii) Acylation: It is carried on by the action of following acylating agents:

(a) *Acyl halides:*

$$Ar-H+R-\overset{\overset{O}{||}}{C}-Cl \xrightarrow{AlCl_2} Ar-\overset{\overset{O}{||}}{C}-R+HCl$$

(b) *Acid anhydrides:*

$$Ar-H+(RCO)_2O \xrightarrow{AlCl_2} Ar-\overset{\overset{O}{||}}{C}-R+R-COOH$$

Mechanism:

$$R-\underset{\underset{O}{||}}{C}-Cl+AlCl_3 \longrightarrow R-\underset{\underset{O}{||}}{\overset{\delta_+}{C}}..Cl..\overset{\delta_-}{Al}Cl_3 \longrightarrow R-\underset{\underset{O}{||}}{\overset{+}{C}}+AlCl_4$$

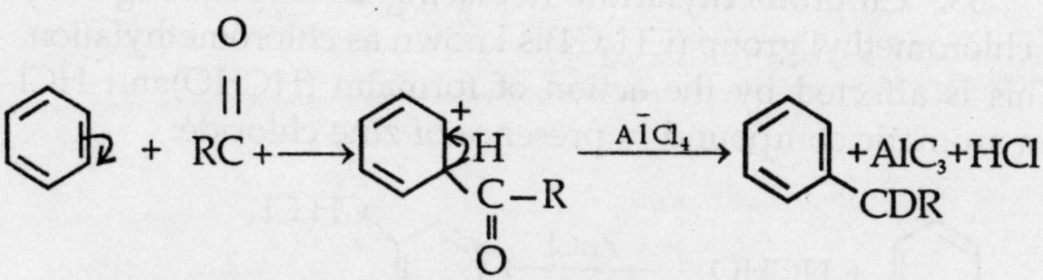

Limitations of Friedel-Crafts reaction:

(A) We cannot check poly substitution

(B) It is not possible to introduce long n-alkyl groups (such an *n*-propyl or butyle) because *n*-alkyl (primary) carbonium ion, being less stable, rearranges to secondary or tertiary.

(C) The presence of *m*-directing (*deactivating*) group in the aromatic ring makes Friedel-Crafts reaction difficult.

32. Chlorination in side chain of toluene. In absence of catalyst, at boiling temperature and in presence of sunlight, chlorine reacts with toluene to bring about the substitution in methyl group., producing benzyl chloride benzal chloride and benzotrichloride successively.

Mechanism:

$$Cl_2 \xrightarrow[\text{[heat]}]{hv} Cl. + Cl.$$

$$C_6H_5-CH_3+Cl. \longrightarrow C_6H_5-CH_2+HCl$$

$$C_6H_5-CH_2+Cl-Cl \longrightarrow \boxed{C_6H_5-CH_2-Cl} + \cdot Cl$$

$$C_6H_5-CH_2-Cl+.Cl \longrightarrow C_6H_5--CH-Cl+HCl$$

$$C_6H_5-CH-Cl+Cl \longrightarrow \boxed{\begin{matrix} C_6H_5-CH-Cl \\ | \\ Cl \end{matrix}} + \cdot Cl$$

$$C_6H_5-CHCl_2+\cdot Cl \longrightarrow C_6H_5-CCl_2+HCl$$

$$C_6H_5-CCl_2+Cl-Cl \longrightarrow \boxed{C_6H_5-CCl_3} + \cdot Cl \text{ etc.}$$

33. Chloromethylation: Replacing an aryl hydrogen by a chloromethyl group (CH_2C1)is known as chloromethylation. This is affected by the action of formalin (HCHO)and HCl on aromatic compound in presence of zine chloride:

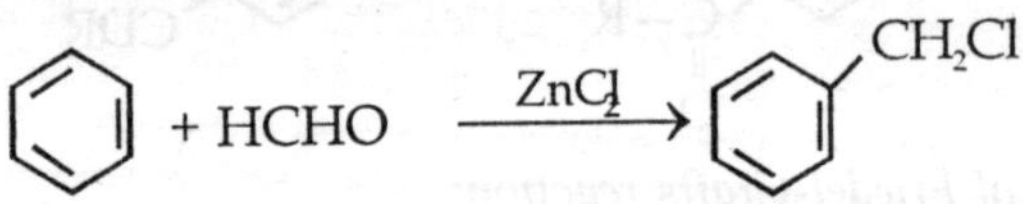

Mechanism:

$$2\,HCHO + 2\,HCl + ZnCl_2 \rightleftharpoons 2\,\overset{+}{C}H_2OH + \overset{-}{Zn}\,Cl_4$$

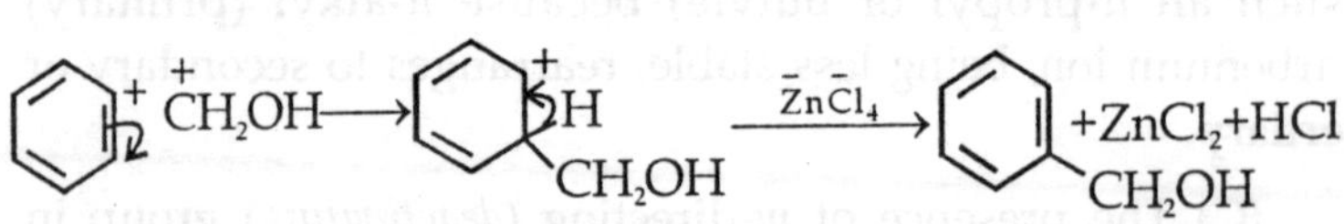

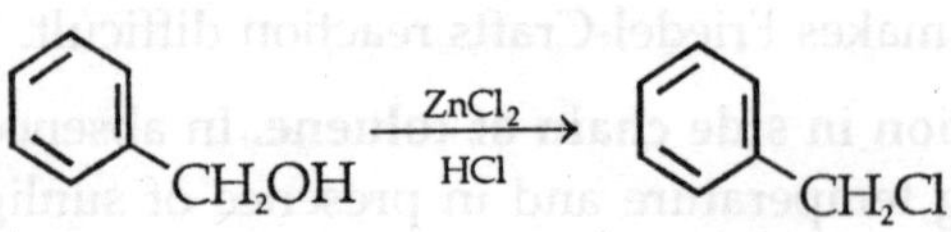

33. Rearrangement of diazoaminobenzene to p-aminoazobenzene: In presence of a mineral acid, diazoaminobenzene rearranges to form *p*-aminoazobenzene.

$$C_6H_5-N=N-NH-C_6H_5 \xrightarrow[40^\circ]{HCl} H_2N-C_6F_4-N=N-C_6H_5$$

Mechanism:

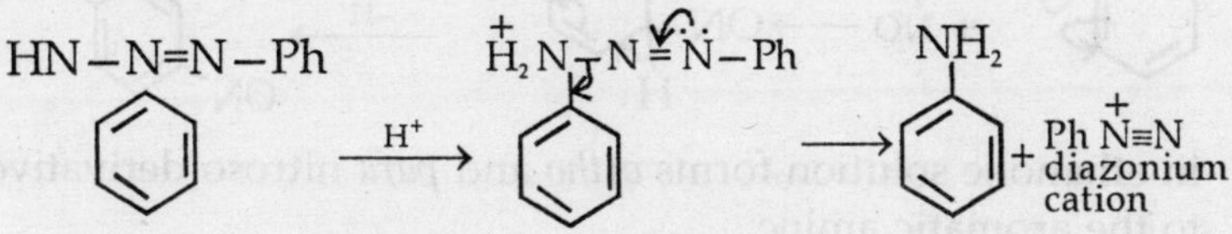

The reactive diazonium cation is an electrophile and can act as host for π electrons of the aromatic ring.

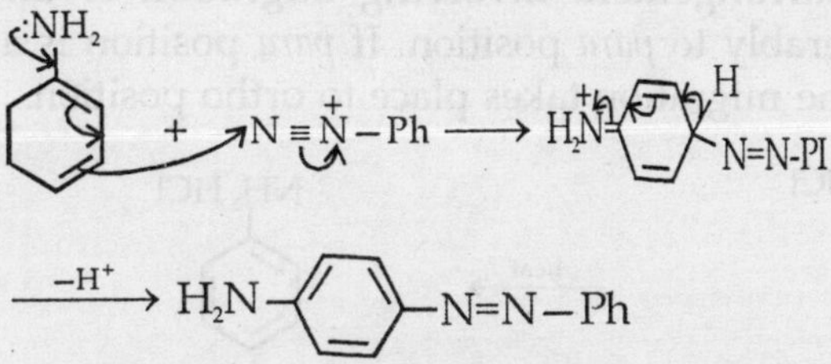

34. Fischer-Hepp rearrangement: N-nitroso derivatives of aromatic amines undergo rearrangement in presence of conc. HCl

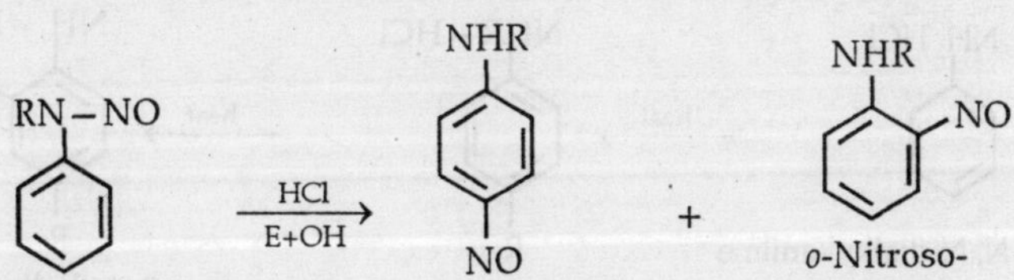

N-Alkyl, N-Nitroso aniline — *P*-Nitroso-N-Alkylaniline — *o*-Nitroso-N-alkylaniline

Mechanism:

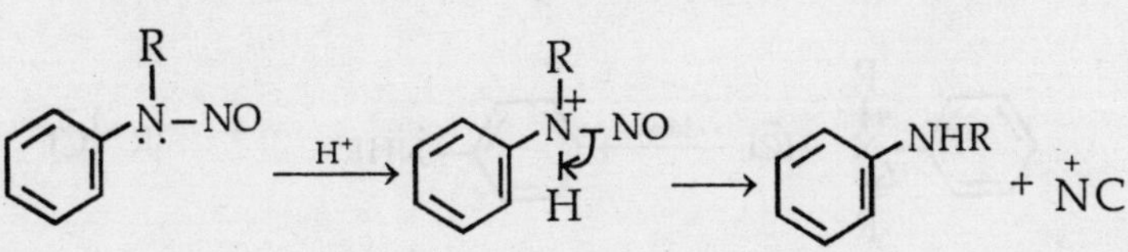

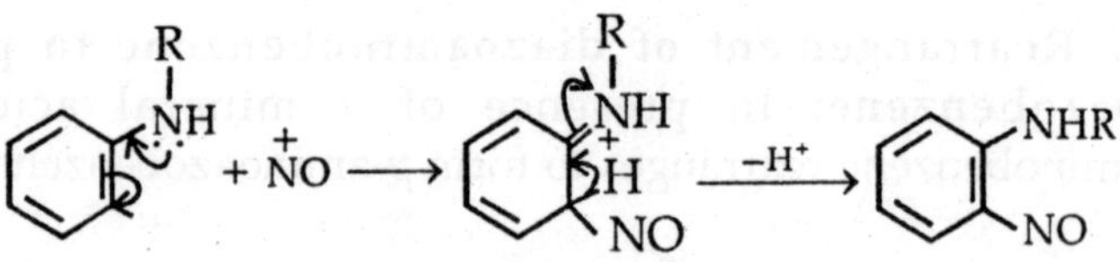

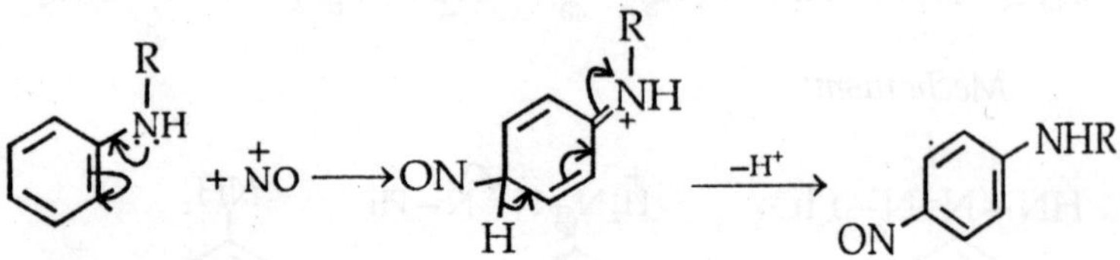

in ethanolic solution forms *ortho* and *para* nitroso derivatives to the aromatic amine.

35. Hofmann-Martius rearrangement: Hydrochlorides of mono- and dialkyl anilines on strong heating (260°-275°) undergo rearrangement involving migration of the alkyl group preferably to *para* position. If *para* position is already occupied, the migration takes place to ortho position.

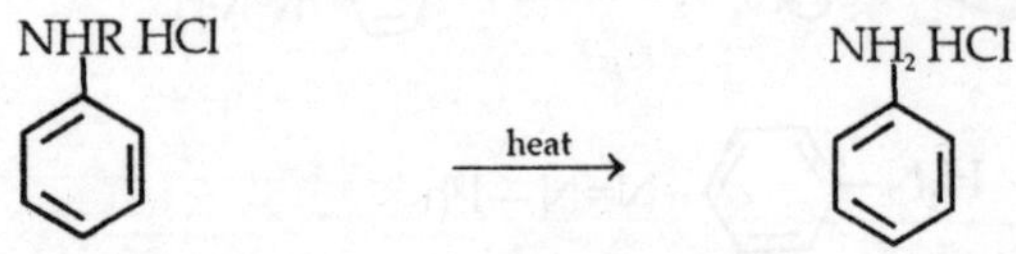

N-alkyl-aniline hydrochloride → p-alkyl-aniline hydrochloride

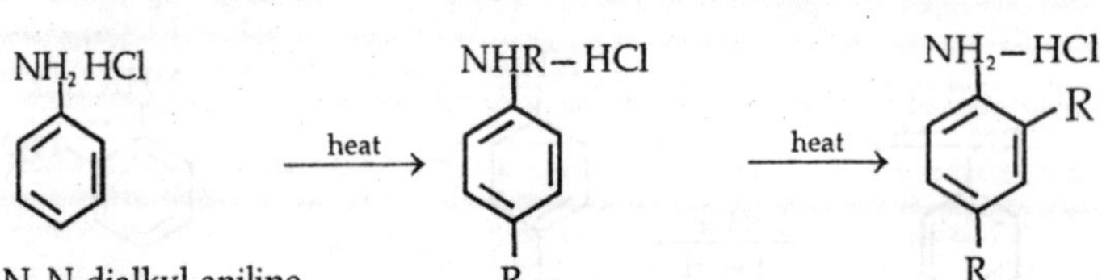

N, N-dialkyl-aniline hydrochloride → P-alkyl, N-alkyl-aniline hydrochloride → 2, 4-dialkyl-aniline hydrochloride

Mechanism:

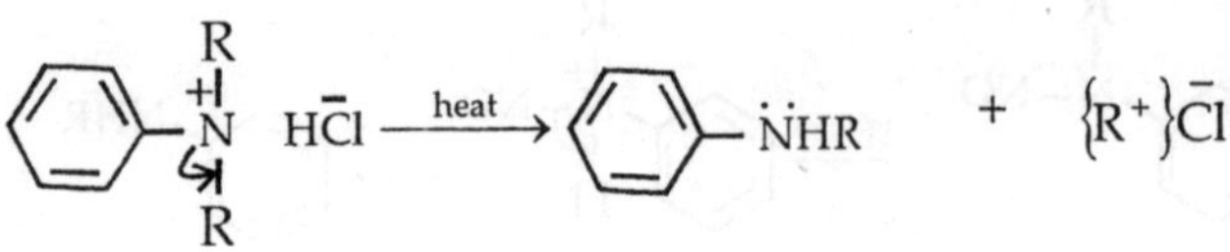

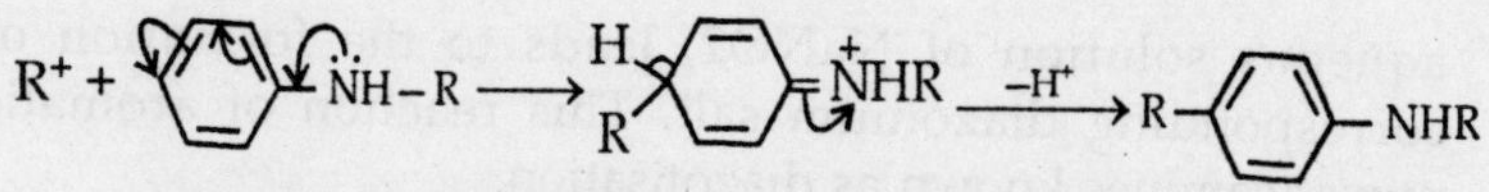

36. Rearrangement of phenyl hydroxylamine to *p*-aminophenol. In dilute acid solution, phenyl hydrxoylamine rearranges to form p-aminophenol.

$$C_6H_9-NHOH \xrightarrow[HCl]{dil} p-C_6H_4\begin{smallmatrix}OH\\NH_2\end{smallmatrix}$$

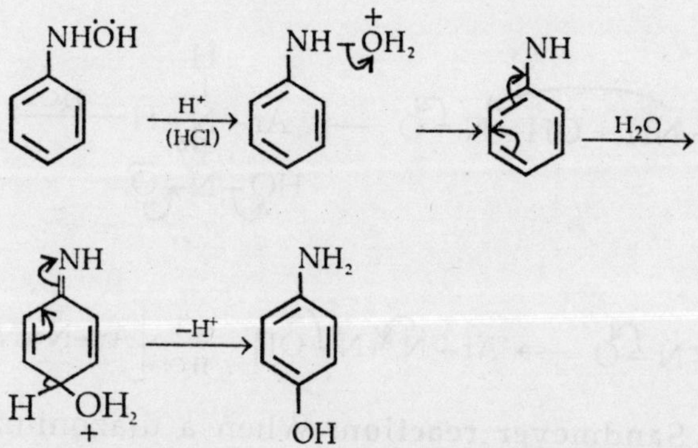

37. Benzidine rearrangement: When hydrazobenzene is warmed with dil HCl undergoes rearrangement and forms *p, p′* diaminodiphenyl (benzidine).

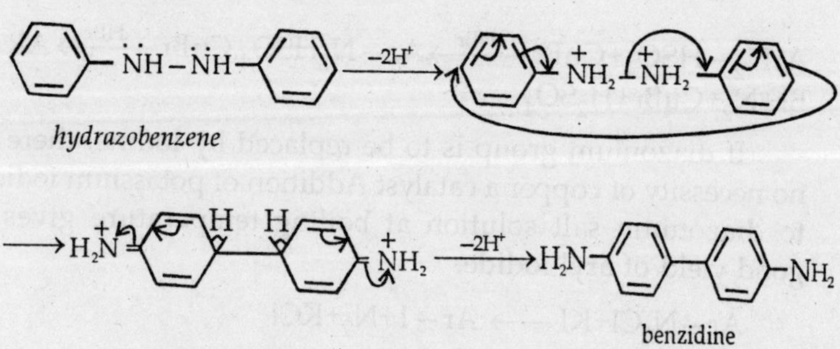

38. Diazotisation: Ice-cold solution of an aromatic *primary* amine in dil . HCl, when treated with ice cold aqueous solution of NaNO2, leads to the formation of corresponding diazonium salt. This reaction of aromatic *primary* amines known as diazotisation.

$$Ar-NH_2+HCl \longrightarrow Ar-NH_2HCl$$

$$NaNO_2+HCl \xrightarrow[O^\circ-5^\circ]{} HNO_2+NACl$$

$$Ar-NH_2.HCl+HNO_2 \longrightarrow Ar-\overset{+}{N}_2\overset{-}{Cl}+2H_2O$$

diazonium salt

Mechanism:

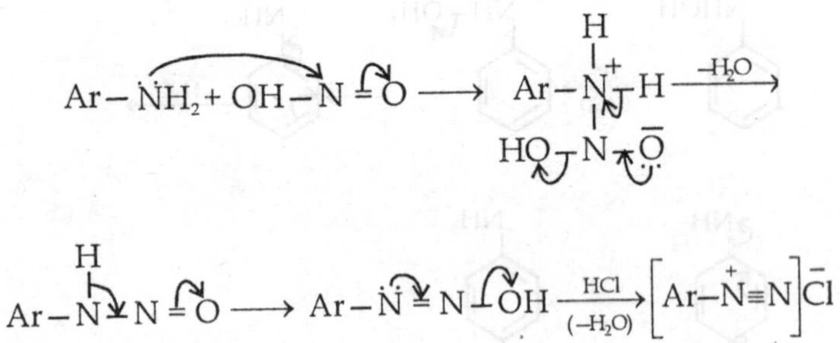

39. Sandmeyer reaction: When a diazonium salt solution is added to cuprous halide dissolved in the corresponding halogen acid, a double salt is formed which when warmed decomposes to form the aromatic halide.

$$Ar-N_2Cl+CuCl \xrightarrow{HCl} Ar-N_2Cl.CuCl \longrightarrow Ar-Cl+N_2+CuCl$$

$$ArN_2-HSO_4+CuBr \xrightarrow{HBr} Ar-N_2HSO_4.CuBr \xrightarrow{HBr} AR-Br+N_2+CuBr+H_2SO_4$$

If diazonium group is to be replaced by Iodine, there is no necessity of copper a catalyst Addition of potassium iodide to diazonium salt solution at boiling temperature gives a good yield of aryl iodide.

$$Ar-N_2Cl+KI \longrightarrow Ar-1+N_2+KCl$$

Also, this reaction is used for replacement of the diazonium group by CN using CuCN/KCN.

$$Ar-N_2Cl+KCN \xrightarrow{CuCN} Ar-CN+KCl+N_2$$

Mechanism:

$$Ar\overset{+}{N}_2\overset{-}{Cl} \longrightarrow Ar-\overset{+}{N}_2+\overset{-}{Cl}$$

$$\overset{-}{Cl}+CuCl_2 \longrightarrow \overset{-}{Cu}Cl_2$$

$$Ar+\overset{+}{N}_2+\overset{-}{Cu}Cl_2 \xrightarrow{slow} \underset{\text{(complex)}}{Ar-N_2CuCl_2}$$

$$Ar-N_2CuCl_2 \xrightarrow{\text{decomposit ion}} Ar-Cl+N_2+CuCl$$

Also,

$$Ar-N_2CuCl_2 \xrightarrow[\text{(fast)}]{Ar\overset{+}{N}_2\overset{-}{Cl}} [(ArN_2)_2CuCl_2)\overset{+}{]}\ C\overset{-}{l}$$

$$\Big\downarrow \text{Decomposition}$$

$$2Ar-Cl+2N_2+CuCl$$

40. Gomberg reaction: Formation of a diaryl is when an alkaline solution of diazonium salt is treated with an aromatic compound *p*-bromo-Benzediazonium chloride, for example, when treated with benzene in NaOH forms *p*-bromodiphenyl.

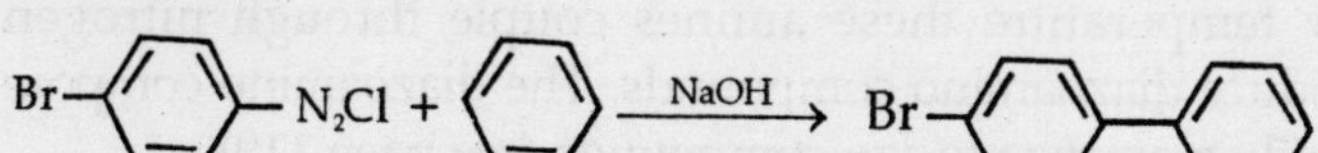

Whatever is the nature of the substituent in the ring of the aromatic compound, substitution occur at ortho or para position w.r.t the position of substituent.

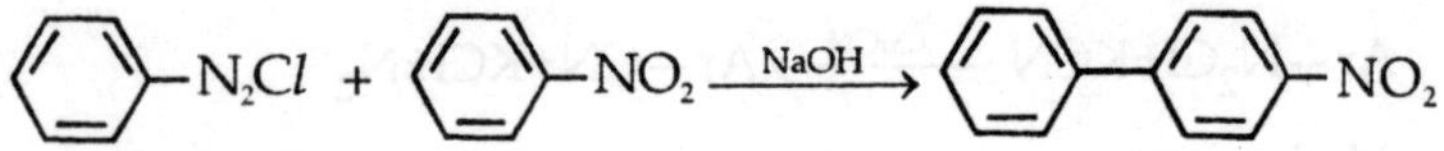

Mechanism:

$$Ar-N_2-Cl \xrightarrow{NaOH} Ar-N_2-OH+NaCl$$

$$Ar-N_2-OH \longrightarrow Ar + N_2 + \cdot OH$$

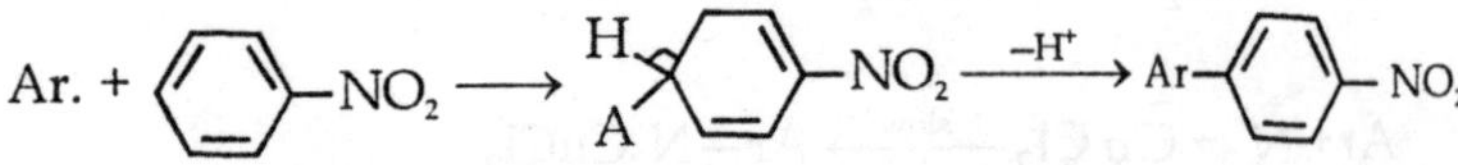

$$H. + OH \longrightarrow H_2O$$

41. Coupling reactions of diazonium compounds: Azo-compounds formed by the action of diazonium salts on phenols and aromatic amines is known as coupling reaction. The reaction requires alkaline, neutral of faintly acidic medium depending upon the nature of the reactants.

(a) *Coupling with phenols*—It occurs in faintly alkaline medium.

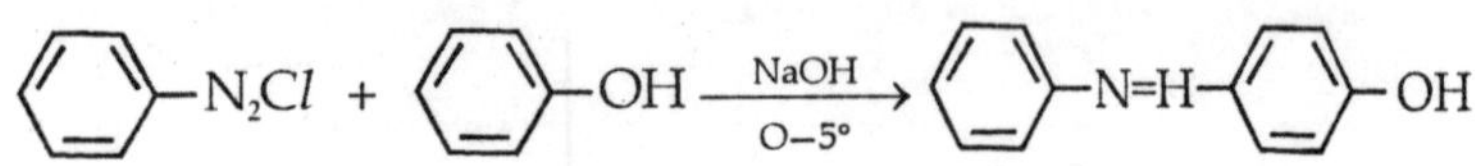

If the *para* position is occupied, coupling occurs at *ortho*.

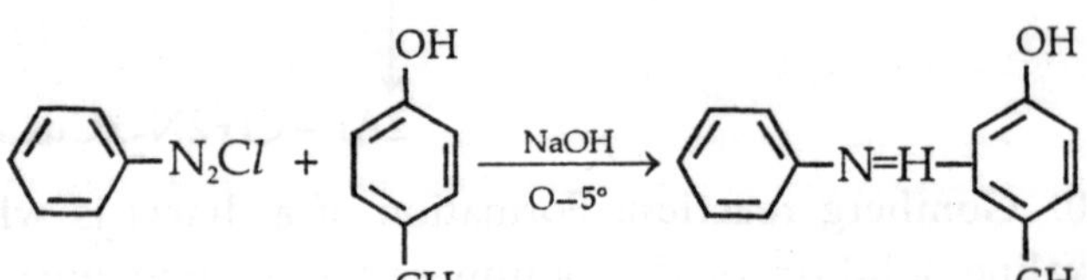

benzene-*azo-p*-cresol

(b) *Coupling with primary and secondary amines*—At very low temperature these amines couple through nitrogen to produce diazoamino compounds. The diazoamino compounds readily rearrange to azo compounds (see page 119).

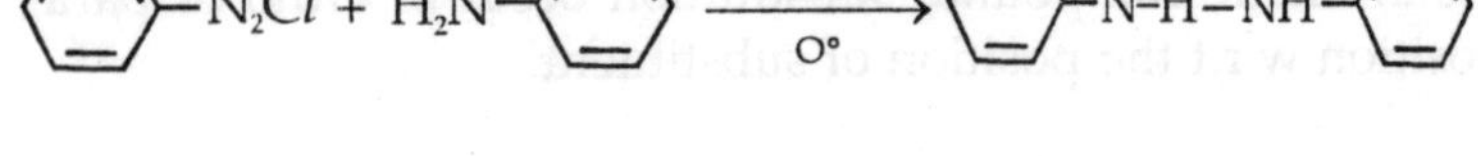

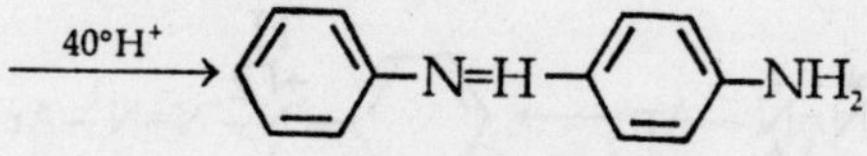

P-aminoazobenzene

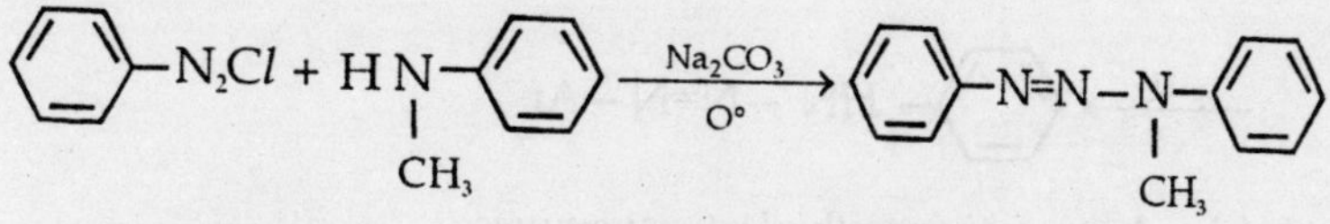

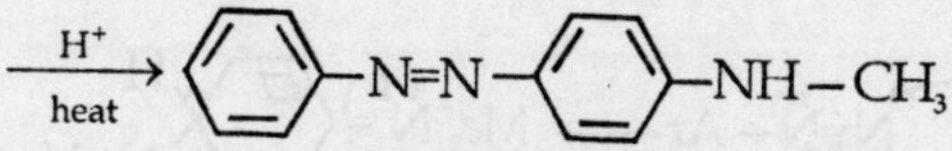

P-N-methyl azobenzene

(C) *Coupling with tertiary amines*—In this case intermediate formation of diazoamino compound is not possible .Amiono-azocompound is directly formed in presence of sodium acetate buffer solution.

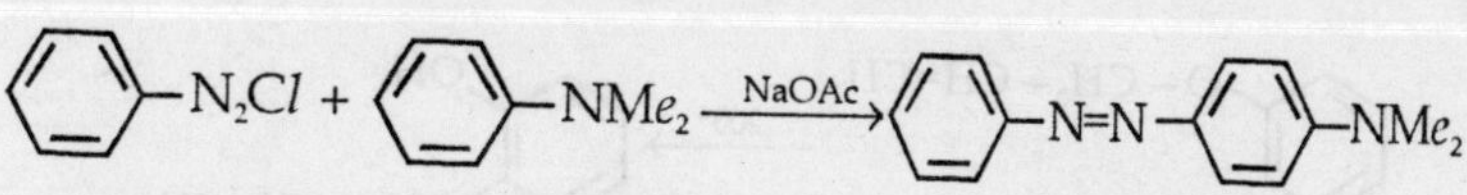

p-N dimethylazobenzene

Mechanism:

(a) *Azo-coupling with phenol:*

$$[Ar-N\equiv N]^{+}\bar{C}l \longrightarrow Ar-N\equiv N + \bar{C}l$$

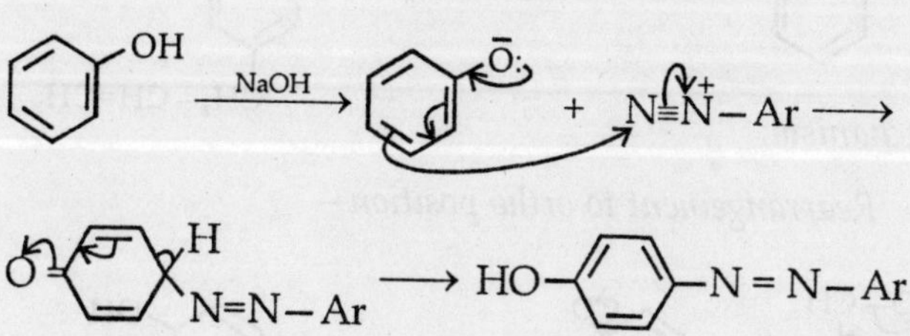

(b) *N-azo coupling with primary or secondary amines:*

$$[Ar-N\equiv N]^{+}\ \bar{C}l \longrightarrow Ar-\overset{+}{N}\equiv N + \bar{C}l$$

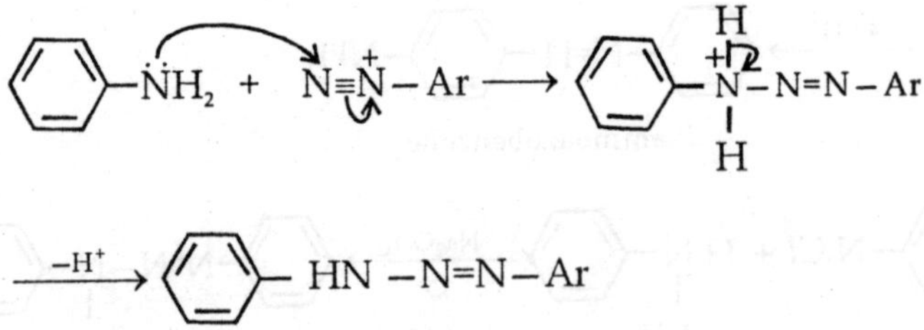

(c) *Azo-couting with plertiary amines:*

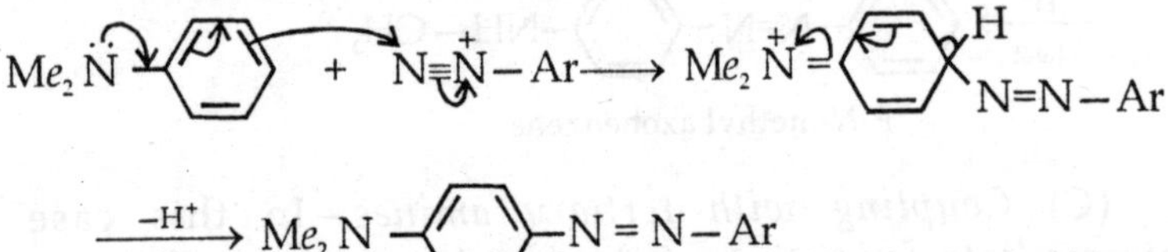

42. Claisen rearrangement: When allyl phenyl ether is heated to 200°; it rearranges to *o*-allyl-phenol.

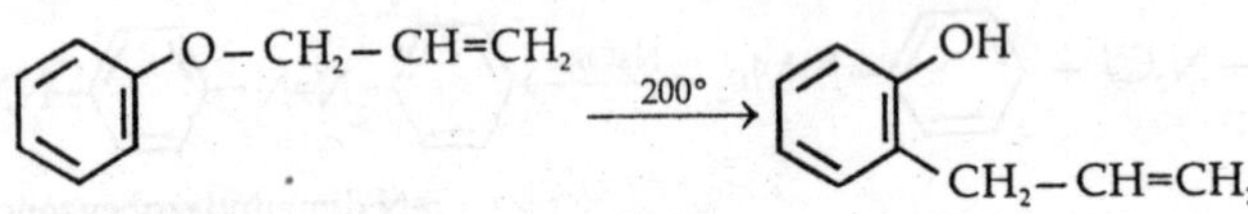

If the ortho positions of the ring are occupied, the allyl group transfers to *para* position:

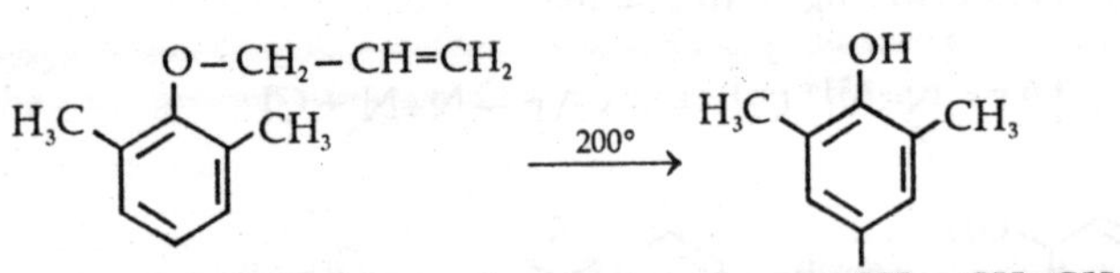

Mechanism:

(i) *Rearrangement to ortho position –*

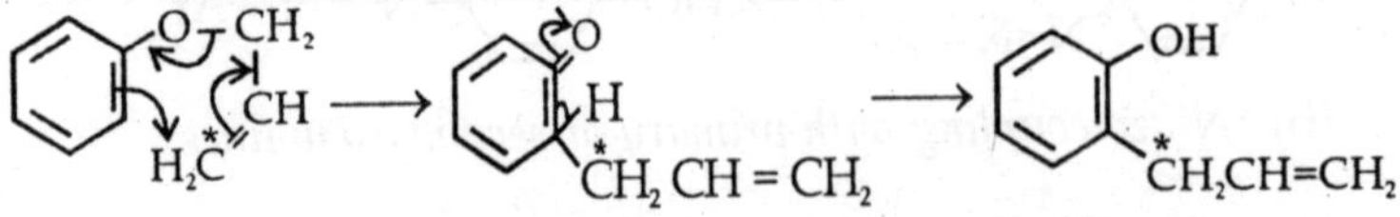

(ii) *Rearrangement to para position –*

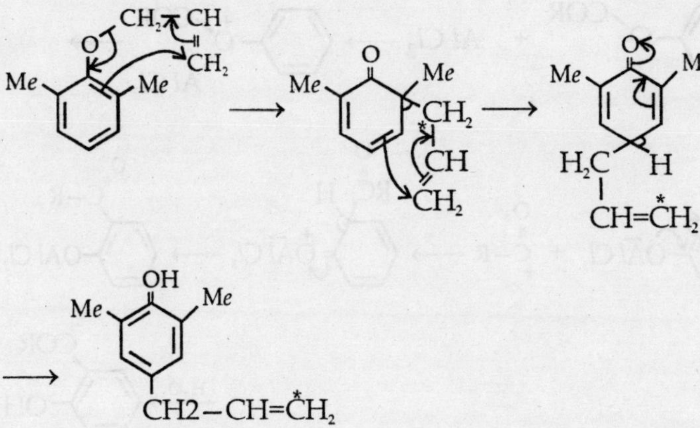

43. **Fries rearrangement:** When phenyl easters are treated with anhydrous $AlCl_3$ it undergoes rearrangement to form *o*-and *p*-hydroxy phenyl alkyl ketones. At low temperature (below 60°), formation

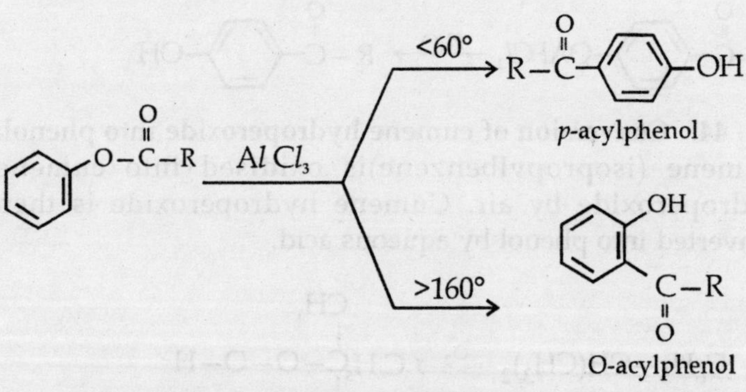

of *p*−isomer is favoured, whereas at temperatures above 160°, the main product is *o*-isomer.

Mechanism: The rearrangement involves formation of acylium cation ($R\overset{+}{C}O$), which brings about electrophilic substitution in the ring.

* Labelling of the carbon atom has been done to indicate the inversion of the allyl group in this rearrangement.

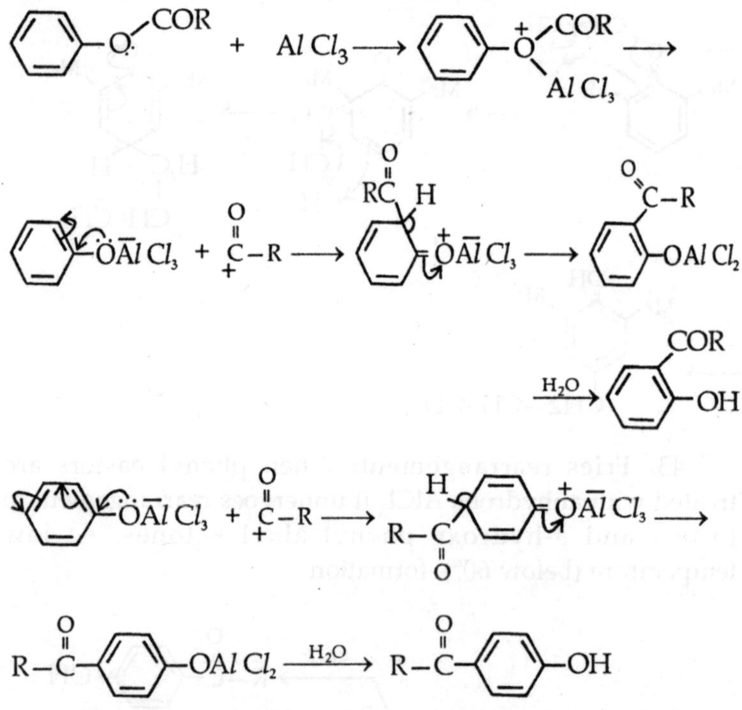

44. Conversion of cumene hydroperoxide into phenol: Cumene (isopropylbenzene)is oxidised into cumene hydroperoxide by air. Cumene hydroperoxide is then converted into phenol by aqueous acid.

$$C_6H_5-CH(CH_3)_2 \xrightarrow{O_2} C_6H_5\overset{\displaystyle CH_3}{\underset{\displaystyle CH_3}{\overset{|}{\underset{|}{C}}}}-O-O-H$$

$$\xrightarrow[H+]{H_2O} C_6H_5OH+CH_3COCH_3$$

Mechanism:

Cumene hydroperoxide is converted to phenol which involves a rearrangement in which a phenyl group migrates to the electron deficient oxygen atom by 1, 2-shift. Following steps are involved.

$$\text{I} \quad (CH_3)_2\underset{\underset{Ph}{|}}{C}-O-\ddot{O}H + \overset{+}{H} \longrightarrow (CH_3)_2\underset{\underset{Ph}{|}}{C}-O-O\overset{+}{H}_2$$

$$\text{II} \quad (CH_3)_2\underset{\underset{Ph}{|}}{C}-O-\overset{+}{O}H_2 \longrightarrow (CH_3)_2\underset{\underset{Ph}{|}}{C}-\overset{+}{O}+H_2O$$

$$\text{III} \quad (CH_3)_2\underset{\underset{Ph}{|}}{C}-\overset{+}{O} \xrightarrow{\text{1, 2-shift}} (CH_3)_2\overset{+}{C}-O-Ph$$

$$\text{IV} \quad (CH_3)_2\overset{+}{C}-O-Ph+H_2O \longrightarrow (CH_3)_2\overset{\overset{+}{O}H_2}{\overset{|}{C}}-O-Ph$$

$$\Downarrow$$

$$(CH_3)_2\overset{OH}{\overset{|}{C}}-O-Ph+H^+$$

$$\text{V} \quad (CH_3)_2\overset{:OH}{\overset{|}{C}}-O-PH \xrightarrow{H^+} (CH_3)_2C=\overset{+}{O}H+Ph-\bar{O}$$

$$\longrightarrow (CH_3)_2C=O+Ph+OH$$

45. Kolbe-Schmitt reaction: When sodium phenoxide is treated at 120°-140° with carbon dioxide (CO_2) under pressure of 100 *lbs.p.s.i.* and the product hydrolysed, salicylic acid (*o*-hydroxy benzoic acid) along with small amount of its *p*-isomer is formed. At temperature above 140 is *p*-isomer the main product is p-isomer.

$$C_6H_5-ONa+CO_2 \xrightarrow[\text{(Pressure)}]{120-140^\circ} C_6H_4\begin{cases}OH\\COONa\end{cases} \xrightarrow{HCl} C_6H_4\begin{cases}OH\\COOH\end{cases}$$

Mechanism;

$$C_6H_5OH+NaOH \longrightarrow C_6H_5\bar{O}:N\overset{+}{a}+H_2O$$

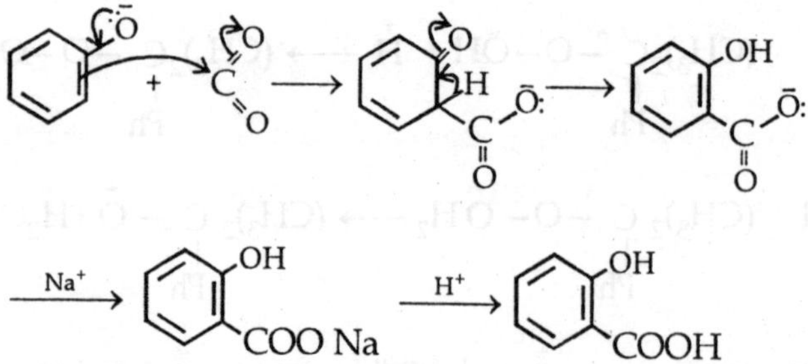

Note: H^+ shift takes place towards phenoxy oxygen and not to the carboxylate oxygen because $CO\bar{O}$ is more resonance-stabilised.

46. Reimer-Tiemann reacton: When an alkaline solution of phenol is heated at 60° with chloroform and the product hydrolysed, a mixture of *o*–and *p*-hydroxybenzaldehydes is formed.

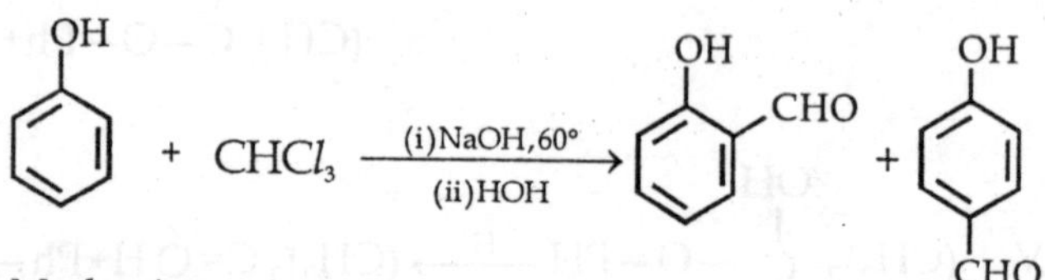

Mechanims:

$$CHCl_3 + :\bar{O}H \longrightarrow H_2O + \bar{C}Cl_2$$

$$\bar{C}Cl_2 \longrightarrow :\bar{C}H + :CCl_2$$

dichlorocarbene

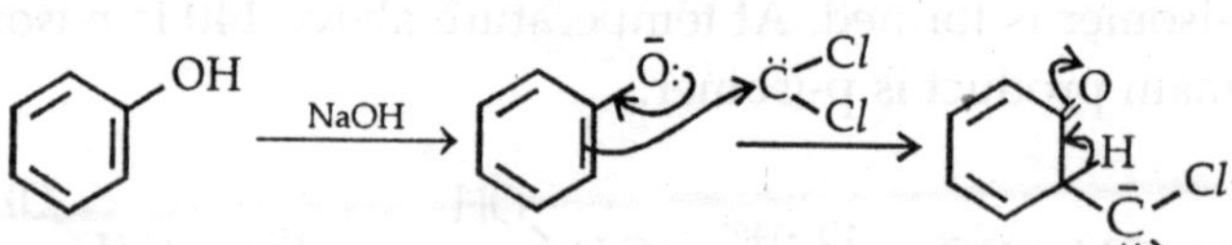

Phenoxy anion

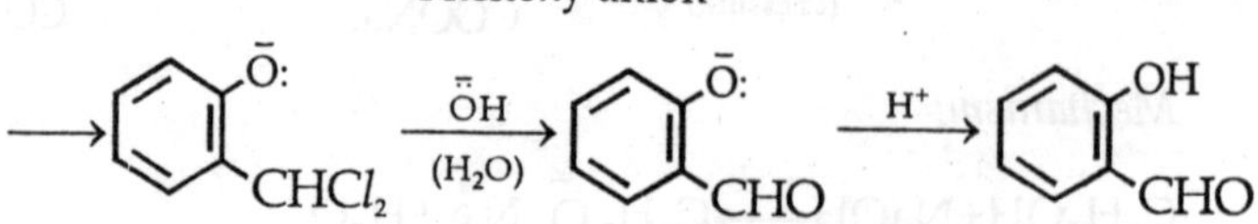

If carbon tetrachloride (CCl_4) is used in place of chloroform, the product is a phenolic acid.

47. Benzoin condensation: When an aromatic aldehyde is refluxed with aqueous ethanolic potassium hyroxide, two molecules of the aldehyde condense to form a benzoin.

$$2Ar-CHO \xrightarrow[\text{aq EtOH}]{\text{KCN}} Ar-\underset{\underset{OH}{|}}{CH}-\underset{\underset{O}{||}}{C}-Ar$$

Mechanism:

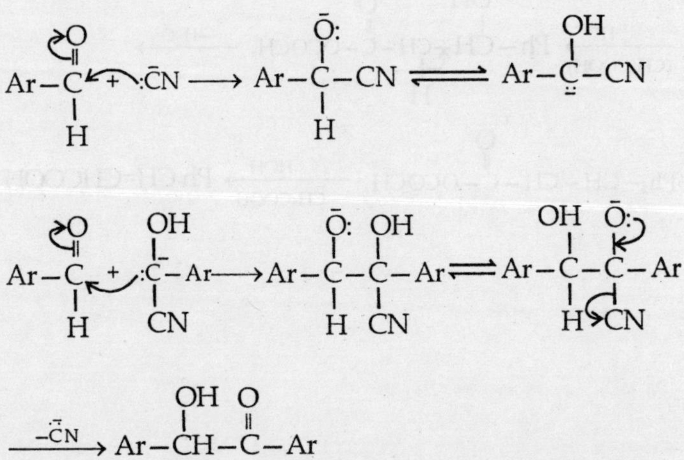

48. Perkin reaction: It is an aldol-type condensation between an aromatic aldehyde and the anhydride of an aliphatic acid must have minimum of two α-hydrogen atoms, in presence of sodium salt of the same aliphatic acid. The product of condensation is a β-arylacrylic acid.

For example—Benzaldehyde when heated at 170° with acetic anhydride and sodium acetate gives cinnamic acid.

$$C_6H_5-CHO+(CH_3CO)_2O \xrightarrow[170^\circ]{CH_3COONa} C_6H_5-CH=CH-COOH$$

Mechanism:

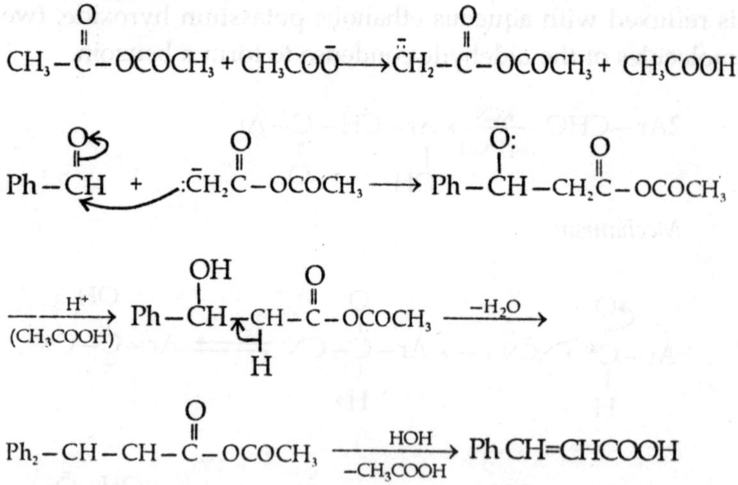

Chapter 9

ACIDS AND BASES

ACIDS

Definition

An acid is a compound (HA)which is capable of donating a proton to the solvent molecule, usually water:

$$HA + H_2\ddot{O}: \rightleftharpoons \overset{-}{A} + H_3\overset{+}{O} \qquad .....1$$

The anion ($\overset{-}{A}$:)left after loss of proton should be more stabilised than the compound (HA) in order to stop the reversible reaction otherwise the proton-loss from HA will not be feasible.

For example, R-COOH (*a carboxylic acid*) is an acid because the carboxylate anion (R-CO$\overset{+}{\ddot{O}}$:) left after proton-loss is more stabilised* than R-COOH itself.

---

* Carboxylate anion is a hybrid of following two resonating structures of *equal energy*:

$$R-\overset{\overset{O}{\parallel}}{C}-\overset{-}{\ddot{O}}: \quad \leftarrow\rightarrow \quad R-\overset{\overset{\overset{-}{\ddot{O}}:}{\parallel}}{C}=O$$

Whereas, carboxylic acid is a hybrid of following two resonating structures of *unequal energy*:

$$R-\overset{\overset{O}{\parallel}}{C}-\ddot{O}H \quad \leftarrow\rightarrow \quad R-\overset{\overset{\overset{-}{O}:}{|}}{C}=\overset{+}{\ddot{O}}H$$

$$R-\overset{\overset{\displaystyle O}{\|}}{C}-OH \rightleftharpoons R-\overset{\overset{\displaystyle O}{\|}}{C}-\bar{O}: + \overset{+}{H}$$

Measure of acidic strength

By applying *law of mass action* to the dissociation of acid HA in water described in Eq. (1), a relation between concentration at equilibrium and a constant, called *dissociation constant or acidity constant (Ka)*, may be given as below:

$$Ka = \frac{[\bar{A}][H_3\overset{+}{O}]}{[HA]}$$ concentration of H_2O being taken as constant.

It is convenient to express acidity constant in logarithmic units, *pKa*, which are negative logarithms of acidity constants, *i.e*

$pKa = -\log Ka$

For example *Ka* for acetic acid in water= 1.75×10^{-5} and *pka* $= -\log (1.75 \times 10^{-5}) = 5 \log 10 - \log 1.75 = 4.76$

More is the acidity of a compound, larger is its, dissociation constant *Ka*, and lesser is its, *pka* value.

Factors that affect the acidic strength

Two factors that are mainly responsible for the extent of acidic character in a compound:

(i) The ease with which the proton is released–

The structural features that tend to facilitate the proton-release will make the compound more acidic whereas those creating hindrance will make it less acidic.

(ii) Extent of stabilisation of the anion left after proton loss–

A componnd will be more acidic if the anion formed after proton-loss has a greater stability due to resonance.

A few examples are given below to describe the effect of structural features on acidic strength:

1. Effect of electron-withdrawing chlorine and electronrepelling methyl group on acidic strengths of aliphatic carboxylic acids.

(i) Due to – I effect of Cl, there is increase acidic strength on going from monochloroacetic acid.trichloroacetic acid.

Chlorine in monochloroacetic acid (I), due to its – I effect, attracts electrons from the chain of σ–bonded atoms making the oxygen of OH partially positive-charged. Oxygen with a partial positive charge facilitates proton-release* by holding the bonding electron-pair of O – H bond. Due to stronger-I effect of 2 & 3 chlorine atoms respectivvely in dichloroacetic acid (II) and trichloroacetic acid (III), the intensity of partial positive charge on oxygen of OH increases, and hence increasing the tendency of proton release. Hence, dichloroacetic acid is a stronger acid than monochloroacetic acid and trichloroacetic acid is stronger then dichloroacetic acid.

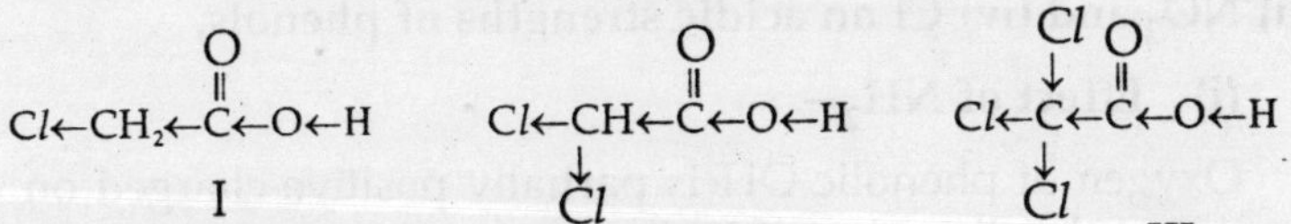

I II III

On a similar ground fluoroacectic acid (FCH_2COOH) is stronger acid than acetic acid and difluoroacetic acid (CHF_2COOH) is stronger than FCH_2COOH, while trifluoroacetic acid (CF_3COOH) is stronger than CHF_2COOH.

(ii) Due to+I effect of methyl group, acidic strength decreases in going from mono-methyl acetic acid to trimethylacetic acid.

In methylacetic acid (IV), the methyl group, due to its+I effect repells the electrons along the chain of σ-bonded atoms creating a partial negative charge on oxgen of OH. In methylacetic acid oxygen with partial negative charge makes the proton release difficult. In dimethylacetic acid (V) and trimethyl acetic acid (VI), due to stronger +I effect of two and three methyl group respectively, the intensity of partial negative charge on oxgen of OH increases, thereby making the proton-release much more difficult. Hence, acetic acid is more stronger than methtylacetic acid and methylacetic acid

* Easier the proton-release, stronger is the acid.

is stronger than dimethylacetic acid, while dimethylacetic acid is stronger than trimethylacetic acid.

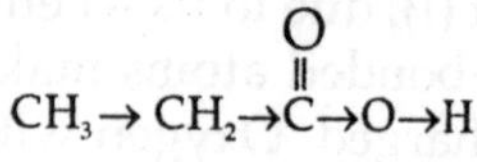

IV

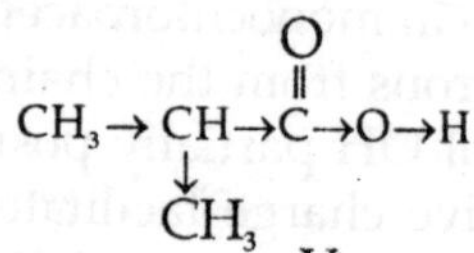

V

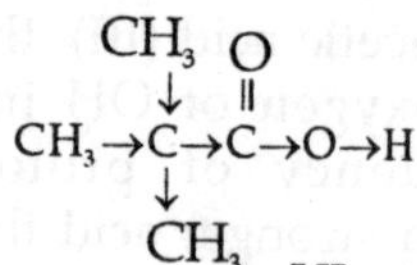

VI

2. Effect of ring-substituents (i) NH_2, (ii) CH_3, (iii) NO_2 and (iv) Cl on acidic strengths of phenols.

(i) Effect of NH_2–

Oxygen of phenolic ÖH is partially positive-charged on account of the diversion of its lone pair of electrons towards the resonance of the aromatic ring (Fig. 24). Due to this positive charge on oxygen, a phenolic OH is able to release H+ and thus acquires the acidic character. Thus, the factors that increase the positive charge on oxygen is respnsible for the increase in the acidic strength of phenol.

In *p*-aminophenol, NH_2 increases the electron density of the ring at its *p*-position due to its+M effect (–I effect is negligible), thereby partially preventing the migration of the lone pair of electrons of oxygen of ÖH towards the ring. So there less positive charge on oxygen atom and proton-release from OH is more difficult in *p*-aminophenol than in phenol. Hence p-aminophenol is a weaker acid than phenol.

In *o*-aminophenol, the predominant –I effect of NH_2 at *ortho* position results in loss of electron density from this position, thereby facilitating the migration of lone pair of electrons on oxygen of OH towards the ring. The positive charge on oxygen atom is therefore greater and the proton-release from ÖH is easier in *o*-aminophenol than in phenol. Hence *o*-aminophenol is a stronger acid than phenol.

The meta position, which is not affected by+M effect of NH in an aromatic ring suffers a loss of electron density due to-effect of this group. Consequently, the migration of the lone pair of electrons on oxygen of OH towards the ring and the positive charge on oxygen in m-aminophenol is greater than that in phenol. m-Aminophenol is thus a stronger acid than phenol.

m-aminophenol is a weaker acid than o- aminophenol because$-$I effect of NH_2 is weaker at meta than at ortho position.

The *pKa* values given below confirm the relative acidic strengths of phenol and the three aminophenols$-$

C_6H_5OH	$H_2N-C_6H_4-OH$	
	o$-$	9.71
9.98	*m*$-$	9.87
	p$-$	10.30

(ii) Effect of CH_3-

Due to +I and +M effects of CH_3, the migration of the lone pair of electrons on oxygen of OH towards the ring in cresols is less than that in phenol. In cresols, the positive charge on oxygen of OH is lesser in phenol which makes cresols a weaker acid. The difference between the acidic strengths of the three cresols is only marginal as is evident from the pKa values given below$-$

C_6H_5-OH	$H_3C-C_6H_4-OH$	
	o$-$	10.28
9.98	*m*$-$	10.08
	p$-$	10.14

(iii) Effect of NO_2-

Due to $-$I and $-$M effects of NO_2, the migration of the lone pair of electrons on oxygen of $\ddot{O}H$ towards the ring is

* Electron-acceptors : alumina, P_2O_5, $ZnCl_2$ etc.

** Proton-donors: H_2SO_4, H_2PO_4 etc.

more in nitrophenols than in phenol. Nitrophenols are stronger acids as positive charge on oxygen of OH is greater in nitrophenols than in phenol.

Since, -M effect of NO_2 does not operate at *meta* position and this position is rendered electron-deficient only by -I effect of NO_2, the electron withdrawl from *meta* position due to NO_2 is less than that from *ortho* and *para*, where besides −I effect, −M effect of NO_2 also operates. The migration of the lone pair of electrons on oxygen of OH towards the ring and positive charge on oxygen are thus less in *m*-nitrophenol than in *o*-and *p*- *m*-Therefore, nitrophenol is a weaker acid than - and *p*-nitrophenols.

Though the combined influence of -I and -M effects of NO_2 in an aromatic ring is greater at *ortho* than at *para* position, yet *o*-nitrophenol is a weaker acid than *p*-nitrophenol. This unusual behaviour of *o*-nitrophenol is due to chelation that hinders the proton-release from OH:

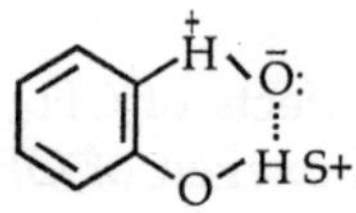

The *pKa* values are given below—

	$O_2N-C_6H_4-OH$	
C_6H_5-OH	*o*—	7.23
9.98	*m*—	8.40
	p—	7.15

(ii) Effect of Cl

Overall electron density of the ring decreases as *Cl* is present in the ring, this is due to its strong -I effect, the weak +M effect being ineffective. The migration of the lone pair of electrons from oxygen of ÖH and hence the positive charge on oxygen being greater in chlorophenols than in phenol, former are stronger acids, This is evident from the *pKa* values given below—

C_6H_5-OH	$Cl-C_6H_4-OH$	
9.98	*o*−	8.48
	m−	9.02
	p−	9.38

The acidic strengths of *o*-, *m*-and *p*-chlorophenols are in decreasing order because -I effect operates in decreasing order from *ortho* to *para*.

3. Effect of ring-substituents (i) NO_2, (ii) OH_2, (iii) Cl and (iv) CH_3 on acidic strengths of aromatic carboxylic acids.

(i) Effect of NO_2-

NO_2, because of its electron-withdrawing influence (-I and −M effects), decreases the electron density of the ring. Hence, the aryl group exerts greater electron-pull* on the O−H bond of COOH in a nitrobenzoic acid than what phenyl group does in benzoic acid. Hence nitrobenzoic acids are stronger acids than benzoic acid.

Since, combined, electron-withdrawing influence of −M and −I effects of NO_2 is maximum at *ortho* position, *o*-nitrobenzoic acid is a stronger acid than its *m*- and *p*- isomers. There is very little difference in the acidic strengths of *m*- and *p*-nitrobenzoic acids which may be due to almost equal electron-withdrawing influence of NO_2 at its *meta* and *para* positions.

The *pKa* values are given below:

C_6H_5COOH	$NO_2-C_6H_4-COOH$	
4.20	*o*-	2.17
	m-	3.35
	p-	3.43

(iii) Effect of OH−

In *p*-hydroxybenzoic acid, OH increases the electron

* The electron-pull due to aryl group on O−H bond facilitates protonrelease from O−H of carboxyl group.

density at its *para* position due to its +M effect (-I effect is negligible at this position), thereby exerting a check on the electron-pull of the aryl group on O−H of COOH. Hence, the proton-release from carboxyl group of *p*-hydroxybenzac acid is difficult and this acid is weaker than benzoic acid, there being no such check in the latter.

In *m*-hydroxybenzoic acid, OH decreases the electron density at its *m*-position due to its −I effect (+M effect is ineffective at this position), thereby increasing the electron-pull of the aryl group on O−H of COOH. Therefore, the proton-release from carboxyl group is easier in *m*-hydroxybenzoic acid than in benzoic acid and the former is a stronger acid.

Increased stabilisation of *o*-hydroxybenzoate anion due to intramolecular hydrogen bonding (chelation) increases the acidic strength of *o*-hydroxybenzoic acid much more than expected. Thus o-Hydroxybenzoic acid is a very strong acid as compared to its isomers as well as to benzoic acid.

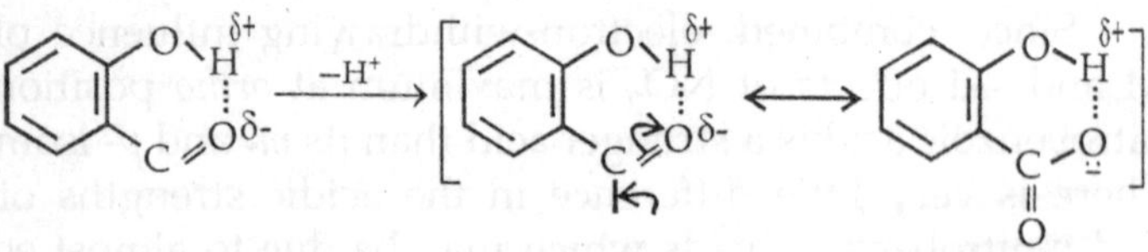

Formation and stabilisation of *o*-hydrxybenzoate anion.

The *pKa* values, given below, confirm the above conclusions.

C_6H_5COOH	$OH-C_6H_4-COOH$	
4.20	*o*-	2.98
	m-	4.08
	p-	4.58

(iii) Effect of Cl−

Chlorine, in chlorobenzoic acid, decreases the electron density of the ring due to its strong -I effect (its +M effect is very weak), thereby increasing the electron-pull of the aryl group on O–H of COOH. So, the proton release from carboxyl group is therefore easier is chlorobenzoic acids than in benzoic acid and the former are stronger acids. This is evident from the *pKa* values given below:

C_6H_5COOH		$Cl-C_6H_4-COOH$
4.20	*o-*	2.94
	m-	3.83
	p-	3.99

The acidic strength of *o-*, m- and *p*-chlorobenzoic acids are in decreasing order, because the -I effect operate in decreasing order from *ortho* to *para.*

(iv) Effect of CH_3-

Due to the combined effect of +M and +I effects CH_3 in toluic acids, increases the electrom density of the ring thereby exerting a check on the electron -pull of the aryl group on O--H bond of COOH. The proton-release from the carboxyl group of toluic acids should thus be difficult and these acids be weak as compared to benzoic acid. The observed pKa values are however, as below–

C_6H_5-COOH		$H_3C-C_6H_4-COOH$
4.20	*o*	-3.91
	m-	4.27
	p-	4.37

As is evident from the *pKa* values, *m*-and *p-* toluic acids are weaker acids than benzoic acid, whereas *o-* toluic acid is stronger. The unusual behaviour of *o*-toluic is due to *steric effect* caused by *close proximity* of CH_3 and COOH groups. Because of this steric effect coplanarity of the ring with COOH is disturbed, thereby inhibiting the resonance of COOH with the ring. Consequently in $COOH_3$, the π electron pair of C=O is entirely free to resonate with lone pair of electrons on oxygen of OH and a positive charge is created at oxygen of O–H. This positive charge exerts electron pull on O–H bond leading to the proton release.

$$-\overset{\displaystyle O}{\overset{\|}{C}}-\ddot{O}-H \longleftrightarrow -\overset{\displaystyle \bar{O}:}{\overset{\|}{C}}=\overset{+}{O}-H$$

Resonating structures of carboxyl group

The combined + M and + I effects of CH_3 at *para* position being greater than its only + I effect at *meta*, the electron-pull on O–H of COOH by aryl group is greater in *m*- toluic acid than in *p*- toluic acid. Former is thus a stronger acid.

BASES

Definition

An organic compound is a base if it is capable of donating an unshared pair of electrons for coordination with proton of the solvent, usually water, to form a *stabilised* cation. This is described in equation (2).

$$\underset{\text{(Base)}}{B:} + H_2O \rightleftharpoons \underset{\text{(Cation)}}{\overset{+}{B}-H} + :\overset{-}{O}H \qquad \text{... (2)}$$

Amines, for example, have an unshare electron pair of nitrogen atom. This electron pair is available for coordination with proton to form a stabilised cation:

$$-\underset{|}{\overset{|}{N}}: + H_2O \rightleftharpoons -\underset{|}{\overset{|+}{N}}-H + :\overset{-}{O}H$$

Thus, amines are basic in aqueous medium.

Measure of basic strength

By applying law of mass action to the equilibrium shown in equation (2), a relation between concentrations at equilibrium and a constant, called equilibrium constant, or basicity constant *(Kb)* may be given below:

$$Kb = \frac{[\overset{+}{B}H][\overset{-}{O}H]}{[B]}$$ concentration of H_2O being

taken as constant

For simplicity basicity constant (*Kb*) is expressed in logarithmic unit (*pKb*). *pKb* is the negative logarithm of *Kb*, *i.e.*,

$$pKb = -\log Kb$$

For example, *Kb* for aniline in water is 4.2×10^{-10}, and its corresponding *pKb* value is $-\log(4.2\times10^{-10})$, *i.e.* 9.38.

Note: Greater the basic strength of a compound, greater the basicity constant, *Kb*, and lesser is its *pKs* value.

Effect of structural features on basic character of nitrogen compounds

If there are such groups or atoms present in the system which due to their electron-withdrawing effect make the nitrogen atom having unshared electron-pair, a centre of decreased electron density, thus hindering the electron - pair on nitrogen atom to coordinate with a proton, decrease the basic character of the compound. Whereas, the groups or atoms enriching the electron density at nitrogen atom due to their electron- repelling effect, facilitate the proton-cooordination and thus make the compound more basic. Some examples are below for illustration—

(i) Comparative basic strengths of prim, sec and tertiary amines:

The number of alkyl group on nitrogen increases from *primary* to *tertiary* amines, *viz*, RNH_2, R_2NH and R_3N. The +I effect of the increasing number of alkyl groups tends to increase protonation. Hence, secondary amines are stronger bases than the primary amines. In tertiary amines, though the +I effect of the three alkyl groups is greater than that of the two alkyl groups in secondary amines, yet due to the *steric hindrance* created by the three alkyl groups, the availability of the lone pair of electrons for proton-coordination and subsequent formation of a stabilised ammonium cation is less than that in even *primary* amines.

So, the tertiary amines are the weakest bases of the three classes of aliphatic amines.

(ii) Amides ($R-CO-NH_2$) are almost non-basic:

The lone pair of electrons on nitrogen atom in an amide is

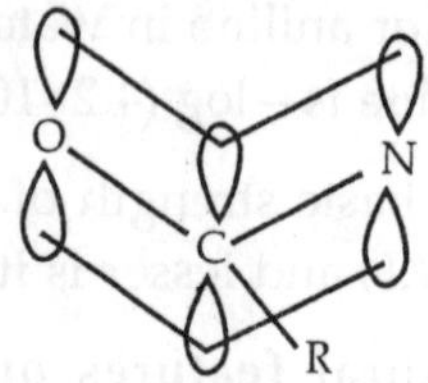

Fig 36 – Delocalisation of the lone pair of electrons on nitrogen in an amide.

conjugated with the the electrons of the carbonyl group (Fig. 36), giving rise to following two resonating structures:

Thus, the lone pair of electrons being engaged in delocalisation is not free for coordination with proton. Moreover, a positive charge on nitrogen atom repels the proton. Hence, NH_2 group does not impart basic character to amides.

(iii) Aniline is a weaker base than aliphatic primary amines:

The lone pair of electrons on nitrogen in aliphatic amines ($R-\ddot{N}H_2$) is localised, whereas in aniline it is delocalised because of its conjugation with the delocalised π system of the aromatic ring (Fig. 37). The availability of the lone pair of electrons for proton-cooordination is thus less in aniline than in alphatic amines So, aniline is a weaker base than aliphatic amines.

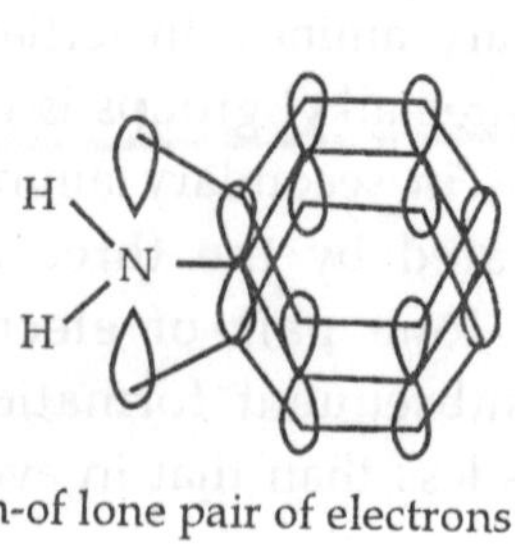

Fig. 37 – Delocalisation-of lone pair of electrons on nitrogen in aniline.

Explanation of the basis of resonance—Protonation in aniline is sufficiently hindered due to (a) the positive charge on nitrogen

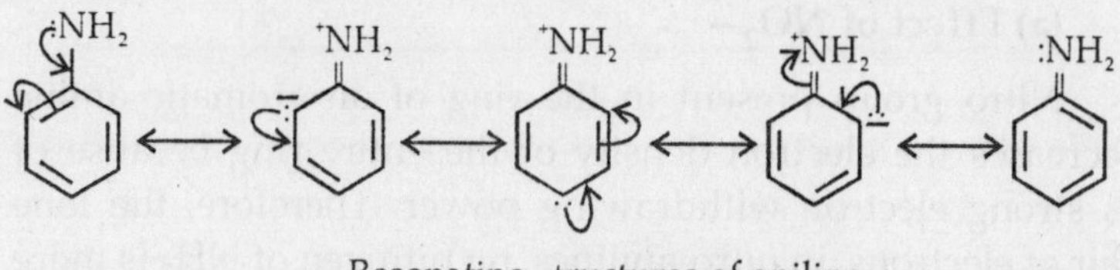

Resonating structures of aniline.

in three of its resonating structures and (b) lesser stabilisation* of anilinium cation($C_6H_5\overset{+}{N}H_3$) than aliline itself. On the other hand in aliphatic amines, there is no such hindrance for protanation. Hence, aniline is a weaker base than aliphatic amines.

(iv) Acetylation markedly reduces the basic strength of aniline.

The lone pair of electrons in aniline on nitrogon is in conjugation with the aromatic ring and so delocalised. In the acetylation product of aniline, $Ph-\ddot{N}H-CO-CH_3$, on the other hand, the lone pair on nitrogen is in conjugation with the aromatic ring as well as with the carbonyl group and is delocalised towards both (Fig. 38) Therefore, as a result of acetylation of aniline, the delocalisation of lone pair of electrons on nitrogen increases, thereby decreasing the availability of this electron pair for

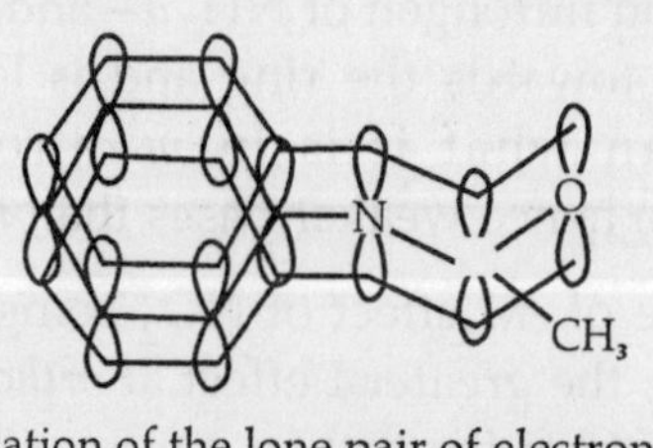

Fig.38—Delocalisation of the lone pair of electrons on nitrongen in acetylation product of aniline (acetanilide).

protonation to such an extent that basic strength of aniline is markedly reduced.

---

* In anilinium cation nitrogen being sp^2 hybridised, has no electron pair in p orbital. The delocalisation of the π electrons of the ring is thus not extended to nitrogen. Hence, the cation is less stabilised than aniline.

(v) Effect of ring-substituents (a) NO_2, (b) Cl, (c) CH_3 and (d) OH on the basic strength of aromatic primary amines:

(a) Effect of NO_2–

Nitro group present in the ring of an aromatic amine decreases the electron density of the entire ring because of its strong electron withdrawing power. Therefore, the lone pair of electrons, in nitro-anilines, on nitrogen of NH_2 is more attracted towards the ring and less available for proton-coordination than what it is in aniline. Nitroanilines are thus weaker bases than aniline. This is evident from the *pKb* values given below:

$C_6H_5-NH_2$	$O_2N-C_6H_4-NH_2$	
9.38	*o*-	14.28
	m-	11.55
	p-	13.02

NO_2 decreases the electron density of the ring due to its-I and -M effects -I effect operates at all the three positions, decreasing from *ortho* to *para*. -M effect, on the other hand, operates only at *ortho* and *para* positions; *meta* remain unaffected. The influence of -M effect NO_2 at *ortho* and *para* positions being greater than the only -I effect at *meta; ortho* and *para* positions are more electron-deficient. Hence, the lone pair of electrons on nitrongen of NH_2 *o*–and *p*–nitroanilines is more diverted towards the ring and is less available for protonation than what it is in *m*-nitroaniline. *o*-and *p*-nitroanilines are hence weaker bases than *m*-nitroaniline.

The influence of -M effect of NO_2 being similar at *ortho* and *para* position, the greater-I effect at *ortho* position makes *o*-nitroaniline a weaker bases than *p*-nitroaniline.

(b) Effect of Cl–

Chlorine, if present in the ring of an aromatic amine, decreases the electron density of the ring due to its strong -I effect, the weak +M effect being ineffective. Therefore, in chloroanilines, the lone pair of electrons on nitrogen of NH_2 is more attracted towards the electron-deficient ring and less available for protonation than what it is in aniline. Chloroanilines are thus weaker bases than aniline. This is evident from the *pKb* values given below:

$C_6H_5-NH_2$		$Cl-C_6H_4-NH_2$
9.38	*o-*	11.36
	m-	10.66
	p-	10.02

The basic strengths of *o*−, *m*− and *p*− chloraoanilines are in decreasing order, as the -I effect operates in decresing order from *ortho* to *para*.

(c) Effect of CH_3-

Methyl group being electron - releasing group on account of its +I and +M effects, increases the electron density of the ring. Therefore, in toluidines, the methyl group prevents the migration of the lone pair of electrons on nitrogen onwards the ring. The lone pair of electron on nitrogen is thus more available for protonation in toluidines than in aniline. Hence toluidines must be stronger bases than aniline. The observed *pKb* values are, however, as below:

$C_6H_5NH_2$		$H_3C-C_6H_4-NH_2$
9.38	*o-*	9.62
	m-	9.33
	p-	9.00

In case of *o*−toluidine, although the inductive and mesomeric effects of CH_3 operate strongly, yet the steric effect due to proximity of CH_3 and NH_2 group prevents the protonation of the lone pair of electrons on nitrogen to such an extent that *o*−toluidine is a weaker base than even aniline.

p−toluidine is a stronger base than *m*−because CH_3

increases the electron density more at *para* position that at *meta* (+ M effect does not operate at *meta* position).

(d) Effect of OH

OH group exerts – I and +M effects on the aromatic ring. The electron density at *ortho* and *para*- positions is increased due to +M – effect being stronger than the – I effect In *o*-and *p*-aminophenols, the migration of lone pair of electrons from nitrogen of NH_2 towards the ring is thus checked (due to electron-rich *ortho* and *para* positions with respect to OH) and aniline. *o*-and *p*-aminophenols are therefore stronger bases than aniline.

The influence of +M effect of OH being similar at *ortho* and *para* positions, the greater -I effect at *ortho* position makes the *o*-aminophenol a weaker base than *p*-aminophenol.

In *m* – aminophenol, the *meta* position with respect to OH suffers a loss of electron density due to -I effect of OH, +M effect being ineffective at this position. So, the lone pair of electrons on nitrogen of NH_2 in *m*-aminophenol is more attracted towards the ring and less available for protonation than what it is in aniline. *m*-aminophenol is therefore a weaker base than aniline.

The *pKb* values, given below, confirm the relative basic strengths of aniline and aminophenols –

$C_6H_5-NH_2$	$HO-C_6H_4-NH_2$	
9.38	*o*-	9.28
	m-	9.83
	p-	8.50

(iv) Comparative basic strengths of aniline, diphenylamine and triphenylamine:

In an aromatic amine, the lone pair of electrons on nitrogen is delocalised towards all such aromatic rings which are attached with nitrogen. Therefore,with increase in number of phenyl groups attached to nitrogen of an amino group., the delocalisation of the lone pair of electrons of nitrogen increases(Fig. 39), thereby decreasing the availability of the lone pair for protonation. As the number of pheryl groups on nitrogen increases from aniline to triphenylamine, the order of basic strength of these amines is –

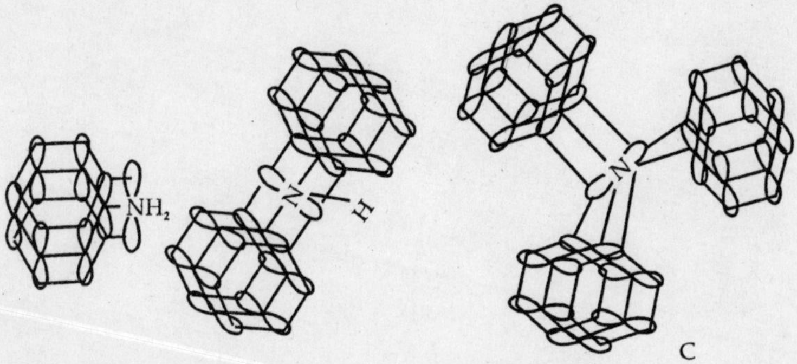

A B C

Fig-39 – Delocalisation of lone pair of electrons on nitrogen in (A) aniline(B) diphenylamine and (C)triphenylamine.